Design Science and Innovation

Series Editor

Amaresh Chakrabarti, Center for Product Design and Manufacturing, Indian Institute of Science, Bengaluru, India

The book series is intended to provide a platform for disseminating knowledge in all areas of design science and innovation, and is intended for all stakeholders in design and innovation, e.g. educators, researchers, practitioners, policy makers and students of design and innovation. With leading international experts as members of its editorial board, the series aims to disseminate knowledge that combines academic rigour and practical relevance in this area of crucial importance to the society.

Koumudi Patil

Toying with the Mind

Eliciting the Cultural Design Process of Bāṇārasī Toy Designers

Springer

Koumudi Patil
Department of Humanities and Social
Sciences
Indian Institute of Technology Kanpur
Kanpur, Uttar Pradesh, India

ISSN 2509-5986 ISSN 2509-5994 (electronic)
Design Science and Innovation
ISBN 978-981-96-9226-2 ISBN 978-981-96-9227-9 (eBook)
https://doi.org/10.1007/978-981-96-9227-9

This Springer imprint is published by the registered company Springer Nature Singapore Pte Ltd.
The registered company address is: 152 Beach Road, #21-01/04 Gateway East, Singapore 189721, Singapore

Foreword

Toys are among the world's oldest known artifacts. They have evolved from sacred and ceremonial objects to ludic playthings, and further to become decorative items, collectable commodities, and museum pieces. In India, the site of the fascinating research reported in *Toying with the Mind*, toy horses, elephants, carts, and rattles are depicted on religious panels dating back to the 2nd Century, testament to a rich tradition of play with both symbolic and ludic purposes.

In this erudite and compassionate study, Koumudi Patil adds to the valuable research on indigenous toys, joining Alan Dundes, David Lancy, Michel Manson, John Roberts, and Jean Pierre Rossie as an indispensable source of data and close observation on toys and toydesigning.

Toying with the Mind continues the groundbreaking studies of handmade toys in *Banārasī* (Varanasi), a holy city on the Ganges river. The work builds on Dr. Patil's doctoral research, for which she won the International Toy Research Association (ITRA) prize in 2014.

Members of that Association will remember with pleasure and wonder the cleverly crafted handmade toys demonstrated by Sudarshan Khanna, the Indian toy designer and educator. Made of paper, wood, or simple available materials, these colorful toys can climb, fly, dance, and make music. Toys (*Khilōnā* in Hindi) are a fundamental part of India's cultural heritage. They bring their users in direct contact with the culture of those who made them.

In recent decades these handmade toys have been discovered by tourists and international museums and crafts markets, placing unprecedented demands on their production and design. Faced with the onslaught of mass-produced Western toys, the toy designers of India, a diffuse community, most of whom are women, have had to adapt to survive, as they strive to balance these threats to their livelihood and their identities.

Toying with the Mind is about the dynamic tension between modernity and traditional culture. In particular, the encroachment of modern, mainly plastic, toys on the community of those who make wooden toys by hand. The book is a rich repository of information on the creation, distribution and uses of handmade *Khilōnās* in *Bānāras*.

Bānārasī toy designers form small groups based on shared skills, proximity, and familiarity. *Kārkhānās* are decentralized, low-tech labs that contain the necessary tools, raw materials, and skills to meet the needs of the toy designers. *Kārkhānās* are spaces for creating and learning. It is here that much of the research, observations, and informal conversations take place.

Toying with the Mind summarizes more than a decade of Dr. Patil's engagement with the community of *Banārasī* toy designers. She visited studio workshops and local shops, and met with toy designers, local customers, and shopkeepers. What stands out to me is Patil's scientific approach and her deep commitment to crafts and the people who create them.

Cultural Design and Modern Design

In his essay 'Toys', the French philosopher and linguist Roland Barthes (1915–1980) contrasts modern plastic toys with those made of wood. Plastic toys are *'made of graceless materials, the product of chemistry, not nature,'* he writes. Barthes laments the '*gradual disappearance of wood, despite its being an ideal material because of its firmness and its softness, and the natural warmth of its touch. ... It is a familiar and poetic substance, which does not sever the child from close contact with the tree, the table, and the floor.*' (Barthes, in *Mythologies*, 1957).

Dr. Patil approaches *Banārasī* toys as a designer and a scientist, aiming to elucidate the conception and processes involved in making *Khilōnās*, *'from their creation through the incremental, assimilative, and disruptive changes they undergo as they are handed down between generations or cultures...'*

She contrasts the traditional cultural design process of the *Banārasī* toy designers with Modern industrial design. Modern design relies on precision, standardization, and mass production. In contrast, cultural designers begin with the unchanging elements and incorporate change as necessary, depending upon local conditions and demands. Cultural designers are distinguished not merely by their skill but by their capacity to improvise amid unpredictable circumstances encountered in situ, sometimes leading to changes in the toy's production or appearance. There is great variety in the appearance and shape of *Khilōnās*, yet Patil notes that they are all recognizable as *Banārasī Khilōnās*'. Her aim is to make the toy designers' tacit knowledge explicit. What knowledge is required to conceive and create a *Khilōnā*. What gives it its recognizable *Bānāras* flavor?

Cultural Design is presented here as equal to, yet different from, modern industrial toy design. Patil is a proponent of the cultural designers, and champions their holistic approach to innovation and education.

Methodology

What is particularly appealing about *Toying with the Mind* is its combination of etic and emic methods. Patil is both a scientific observer and a participant in the toymakers' community.

She makes excellent use of an impressive array of methods to reflect how toymakers, sellers, and buyers think and talk about toys.

In *Toying with the Mind* the *Banārasī* toy designers themselves describe the process of creating *Khilōnās*. In order to understand each person's implicit knowledge of *Khilōnās*, toy designers were asked to sort objects or photos of objects into groups of similar toys. This method is used in psychological research to reveal an individual's cognitive structure regarding the objects in question. Here, the sorting process reveals the *terminology* that people use in describing toys, the *relationships* among toys, such as their proximity or similarity, and *categories* of toys, grouped together.

Toy designers have a rich vocabulary to describe and categorize *Khilōnās* based on their size, means of production, raw material, and decoration. The decoration and colors of *Khilōnās* suggest their purposes. *Khilōnās* of a religious nature were often grouped together, as were those for ludic play, and toys for display. Patrons tend to focus on the toys' potential uses, whether for play, display, or celebration. Shopkeepers, who serve as intermediaries between toy producers and consumers, share a vocabulary with both toy designers and patrons.

From her observations and data, Patil extracts a set of Conventions (rules referring to particular attributes of a plaything, such as all birds have round eyes), Canons (norms in practice, for instance, pecking toys stand on a wooden base), and Beliefs (e.g., playthings impart values to children). These reflect the implicit knowledge shared by *Bānārasī* toy designers and underlie the practice of toy design.

Differentiation, Accommodation, and Resistance

How does change come about in a community with strict standards, traditions and constraints? Creativity and change often occur gradually, with successive iterations approaching an implicit ideal. Change can also occur suddenly, as when an order arrives from beyond the community with requirements that do not conform to the *Banārasī* tradition. The community has developed strategies to manage these potential threats to their identities and livelihoods. They have created distinct categories to differentiate their authentic cultural products from those designed for external consumption. The decoration and painting of *Khilōnās* for non-*Banārasī* patrons that violate *Banārasī* canons, conventions and customs are designed specifically for modern customers, for instance, with pastel colors, as opposed to the vibrant, contrasting colors typical of *Banārasī Khilōnās*. This distinction sets them apart from religious *Khilōnās* and those intended for play or display, maintaining a clear boundary between cultural authenticity and market-driven adaptations.

Some changes have been adapted into *Banārasī* toys. For example, after multiple international requests for jigsaw puzzles, not traditionally played in *Bānāras*, they are now sold to local patrons, as well as tourists. Toy designers have assimilated the Western jigsaw into their repertoire.

What stands out in this monograph is the author's commitment to the cultural design communities, as seen not only in this research but in her personal involvement with the *Banārasī* toymakers. *Toying with the Mind* is not some nostalgic look at a gradually receding craft. It is a plea to see the toy designers of *Banarasi* as they are, and to learn from their practices, social organization, and cultural representation.

Not content merely to describe the *Bānārasī* toy designers and their social organization, Dr. Patil is currently leading a national initiative focused on the development of rural crafts and artisans. Since 2008, she has been an advocate on behalf of craft communities across India, curating a museum of craft, developing creative visualization software for artisans, and conducting training workshops.

The survival of handcrafts and their communities depends on their ability to adapt to changing tourist and export markets while maintaining their toy designing practices and culture, preserving their identities as *Bānārasī* toymakers. In *Toying with the Mind* Patil shows us how this can be done.

Jeffrey Goldstein
Institute for Cultural Inquiry
University of Utrecht
Utrecht, The Netherlands
j.goldstein@planet.nl

Preface

In 2007, the Department of Design, Indian Institute of Technology Kanpur, India, where I teach, received several generous grants from the Ministry of Textiles to support the "modernisation trajectory" of the handmade communities in Uttar Pradesh, India. We were asked to set up a Central Facility Center, which would feature state-of-the-art technologies and machinery for training handmade practitioners to (1) conceptualise contemporary designs, (2) expand the market for which they produce, and (3) record and document their practices on behalf of their communities and thus help ensure that these practices continue in today's modern, capitalist market. In such a program, handmade artefacts embody the essential occupational skills of a pre-industrial era. At the same time, modern design represents a vocation for the gifted, who cater to consumer demands and support the survival of the handmade industry in an unforgiving market. The moral imperative to uplift the marginalized and less privileged cousin of design is the hallmark of such developmental projects.

However, the chasm between the modern and handmade design disciplines is profound. Bruno Latour (2008) argues that objects are no longer merely "made" or "fabricated" (as in the pre-industrial society of handmade communities) but are instead meticulously "designed"—or, as he prefers, "precautionarily designed" (as in the industrial and post-industrial design profession). Modern designers engage in ideation, planning, management, strategy, and negotiation rather than merely "make" an object. On the other hand, the remnants of outdated hand skills scattered around the globe struggle to adapt to a changing world, and the practitioners of these skills continue to labour over the intricacies of form and function in production. The divide between the two disciplines is relentlessly explained by lack: lack of literacy, lack of technology, lack of disruptive thinking, lack of modern design problem-solving techniques, lack of market, lack of awareness of financial equity, lack of innovation, lack of risk-taking capability, lack of formal education, and so on. However, it is most distinctly visible in the economic inequity of its practitioners. The late Muhammad Tughlaq, a *buzurg* or senior brassware practitioner from Morādābād, India, devoted eleven years of his life to the intricate *Nakāśī*, or engraving, of a silver box. Despite receiving the prestigious *Śilpaguru* award, he could not find a patron willing to purchase his silver box for 11 lakhs ($13,166.85 U.S.)—a price equating to

1 lakh ($1196.63 U.S.) per year of his labour. On the other hand, my design student graduates, who are well-versed in the skill of giving prompts to generative AI (Artificial Intelligence), have landed jobs that pay more than 20 lakh per annum on average ($23,928.48 U.S.).

Despite the languishing status of several handmade practices in India, this sector remains the country's second-largest employer after agriculture. The industry exported $120.06 million (U.S.) worth of handmade artefacts in May 2022 and significantly contributed to India's Gross Domestic Product (GDP). This sector is dominated by female professionals, who form 56% of the total population of handmade practitioners—a dramatic shift from the male-centric urban employment scenario. At the 2022 G7 Summit, the Government of India presented exclusive products designed by Indian handmade communities to seven Heads of State. *Gulābī Mīnākārī* brooch and cufflinks, *Marodī Maṭkā, Bulandshahr* crockery, *Dhokrā Nāndī, Aṭṭar*, and black pottery were made by unknown, uneducated, handmade practitioners unaware of the tastes of the foreign elites. No modern Indian designer products were included in this exclusive package. Despite these glorious accomplishments, weavers have committed suicide in large numbers in Andhra Pradesh (refer to the weaver suicides from 1999 to 2001), the entire community of knife makers from Rampur has shifted to other professions in search of livelihood, and the last known *Karīgara* of *Gañjiphā* cards passed away while searching in vain for an apprentice who would receive his skills and thus continue the traditional design practices.

Numerous explanations attempt to decipher the economic and social gaps facing the cultural custodians of India. These range from exploitative intermediaries and administrative indifference to the obstinacy of handmade practitioners, insufficient patronage, limited raw materials, negligible working capital, the disconnect between rural production centers and urban markets, the caste system, the influx of cheap Chinese goods, and the lingering effects of colonial policies. The question of functionality and form no longer exists because, irrespective of them, all handmade goods have arguably become superfluous today. Yet these factors alone do not fully account for the disparity between the acclaim achieved by Master Tughlaq and the career success of my formally educated design students, which primarily hinges on cognition. This bothers me.

Several insightful, in-depth studies have covered the "wicked" nature of handmade problems. Yet the strongly tacit nature of the discipline has impeded our efforts to unravel the complexity of human decision making. How do handmade practitioners contemplate a problem? What distinguishes their problem-solving approach from that of contemporary designers? Are modern designers truly superior to handmade practitioners in solving problems? These questions and uncertainties lured me into studying handmade practices in India, a journey I embarked upon in 2008 and continue to pursue. This study emerges from my reflections on how handmade practitioners think and act, and on the cognitive processes that underpin their creativity and problem-solving methods.

Specifically, this text is the summary of my more than decade-long engagement with the community of *Bāṇārasī* toy designers residing in Varanasi, India. I have approached *Bāṇārasī* toys as a designer attempting to explicate the design decisions of conceiving and making an artefact by erstwhile designers—the handmade practitioners themselves. While this design study includes elements of linguistic anthropology, it does not intend to contribute to historical, sociological, or philosophical approaches. Rather, it contributes to understanding the development of typicality and variation in culturally sensitive designs.

What Should It Be Called?: The Problem of Labelling

Scholarship in design, art, cultural anthropology, and material culture has extensively discussed the role of handmade practitioners in the creative industry and bracketed them in several types of semantic containers ranging from traditional designers, artisans, informal practices, vernacular practitioners, and nonliterate traditions to unselfconscious designers. These practitioners, irrespective of their reflexivity[1] (which closely resembles that of mainstream modern designers), have been consciously dissociated and categorized to differentiate from what we consider modern, contemporary, industrial, innovative, and creative. Somewhere in the history of design, we shifted our goal from creating a new guild of handmade designers (Gropius 1919) to creating cutting-edge studios that have severed ties with their past. The binaries of useful–useless, modern–traditional, handmade–industrial, innovative–imitative, rural–urban, indigenous–global, and so on, as instigated by sociohistorical narratives have marginalized the erstwhile designers to a corner.

> Calling a thing craft, or a practice handicraft, or a person a "traditional" craftsperson without further scrutiny, doesn't simply miss the extent to which terms of naming are woven into productive and circulatory networks; it misses how naming strategies and particular claims to distinction weigh down the ideological conception of labour. (Wilkinson-Weber and DeNicola 2020, 79)

Labels are far from neutral; they encapsulate tangible power dynamics and mould the categories that guide our thoughts and actions. These labels dictate access to resources, which compels individuals to conform to specific classifications if they are to manage effectively their interactions with institutions. A crucial mechanism in this process is reducing a person's multifaceted reality to a single attribute—such as land ownership or literacy. This reductionist approach frequently identifies perceived pathologies or deficiencies that can be isolated and ostensibly remedied through technological interventions (Wood 1985, 357).

According to Escobar (2011), to harness the potential for transformative change, we must rigorously examine how labels function as instruments of power within particular institutional frameworks. This critical understanding can guide the formation of collective political actions to counter the isolating and externally imposed

[1] Donald Schön (1983) presented the concept of the "Reflective practitioner" who explores improvisatory forms of knowledge in practice based on technical rationality or on logical transparency. This aligns very well with practice-based disciplines such as craft, art, or design where the practitioners are constantly reflecting on their own work to improvise or learn further from their own actions.

labelling processes. Through such efforts, we can begin to challenge and redefine the entrenched power dynamics inherent in institutional labelling practices.

For a long time, and specifically while writing this text, I have been grappling with the problem of indexing handmade practitioners in a way that does not treat them as a lesser cousin of design in thought and practice; I have not had much success in my endeavour. With a fixed anthropological lens on these historical vestiges, the process and products of the erstwhile designers are being doled out the poor-cousin treatment. There is an entire politics of how "needs" are mediated by expert discourses. Designers and developmental experts act as brokers, mediating relations between communities, the market, and sometimes even the state and social movements (Mohsini 2010, 2020; Wilkinson-Weber and DeNicola 2020). Hence to retain a semantic respect for the profession, I chose to use the term *cultural design practice* in this book because it credits this profession with the ability to design, cogitate, and innovate, albeit differently from its industrial counterpart. The culture here is not a fixed, rigid practice but what K. G. Subramanyan (1987) would call "living." Living practices are continuities of creative practice that evolve and change in response to context. The arguments made in this text will assert the reflexivity of cultural design practice and reposition handmade practitioners as designers of abilities to solve problems responsibly and innovatively in ways that are equal to, yet different from the abilities of their modern, city-dwelling cousins.

The Ethics of My Participation

Revisiting human intelligence (versus artificial intelligence) and hand transmission (versus machine learning) to look for clues to pedagogies of thinking and making is not mere nostalgia for the past but a search for an appropriate episteme of generational knowledge of the handmade that can inform contemporary debates and practices. From this vantage point, as an ethnographer, I approach the cultural design practice of the *Bāṇārasī* toy designers in two roles: first, as a researcher attempting to learn about the community of practice; and second, as a learner attempting to learn from the practice. I am an observer at the periphery of the former. At the same time, in the latter, I participate as an active learning member of the community and eventually blur the line between the observer and the observed. Here, I can leverage my position as an insider and an outsider in the same community and create an enriching environment in which to develop in-depth descriptions of the situation as it is and as it is being created. The collected evidence is based on observations of existing negotiations and interactions that I specifically triggered in the participation process. For the former, I became a regular visitor at six *Kārkhānās* in the cultural design hubs at *Lakṣā, Khojwā, and Kāśmīrigañj*, as well as at the local market: *Vishwanath Gali* in Varanasi, India. For the latter, as if I were a patron, I commissioned the toy designers for several designs, both familiar and unfamiliar to them, and I paid them adequately for their work. In addition, I held workshops and toy design competitions to fuel innovation and instill pride in hand skills. The winners of one such competition[2] were awarded a cash prize and certificate by the Honourable Prime Minister of India: Shri Narendra Modi. Over the years, my role has transformed from an intruder to an invisible spectator, and today I am a benefactor and have earned trust and acceptance within the community.

Cultural design practitioners in India are often illiterate and skeptical of procedures involving written formalities. Therefore, although all participants gave verbal consent for their participation in this study, the same could not be solicited in writing due to their hesitation to transact any paperwork. Given this limitation, throughout this text, the names of participants have been changed to maintain anonymity. Cultural design practitioners are also reluctant to participate in laboratory procedures, such as a neuroscience study involving equipment that is unfamiliar to them or requires any change in their geographical context. This made it imperative for all discussions and negotiations to be conducted in situ at the *Kārkhānās* themselves. *Kārkhānās* are decentralized, low-tech maker labs that develop contextually within the local ecosystem of technologies and tools, raw materials, and skills to meet the needs of the cultural design communities. They serve as a space for living, learning, and earning a livelihood for their owners (who own the means of production), and they become an open source of generational and intergenerational know-how, drive experimentation to meet the demands, reap the profits, suffer the losses: in short, they act as alternate nano-entrepreneurial models.

Within these boundaries, this text attempts to understand handmade design as a protean practice of learning, thinking, and designing rather than as an unchanging, repetitive, sequential method of manufacturing objects. It locates the role of cultural design in contemporary practices in general, and the modern industrialized design profession in particular.

Acknowledgements

There is an order in my gratitude, but the order is not that of a pecking ladder.

Every contribution to and insight made in this text is merely an interpretation and rephrasing of what the community of handmade practices in general and the

[2] Readers may wish to look at the relevance of such competitions in encouraging cultural communities to take increasing risks in innovating their designs (e.g., in the community of Hopi basket weavers). Teiwes (1996, 162) says: "Hopi basket weavers have an astonishing ambition and drive to refine their traditional craftsmanship by creating the most intricate and complicated designs and inventing new colour combinations in wickerwork. Their inborn artistic skill and taste are astonishing and manifest in their artwork. Part of the incentive for this constant refinement comes from juried Indian art shows that award prizes. I have repeatedly heard that women who had won a prize immediately started on a new piece to be entered in another juried show a few months later, challenging themselves to repeat, or possibly improve on, the excellence of their art. When listening to the words of many women, especially younger women, it becomes clear that they admire those who do the finest work, and as a result, they strive to produce the finest coiled and wicker basketry possible."

Bāṇārasī toy designers in particular already know, practice, and pass on to the next generation. Nothing in this text reveals any concept that the toy designers do not already know explicitly or tacitly. Hence, this work is a tribute to the generational knowledge that thousands of highly skilled practitioners preserved and innovated upon. The generosity of these toy designers in sharing their knowledge, resources, and meagre meals with me has humbled and enriched me. I am indebted to them for their trust in me and the patience with which they have explained the "wickedness" of their context-specific problems and specific methods of resolving them.

A significant part of my understanding, as stated in this text, stems from my doctoral work under the supervision of Prof. Uday Athavankar at the Industrial Design Center at the Indian Institute of Technology Powai, India. My discussions and quarrels with him during my doctoral studies have shaped my perspective on this discourse.

I must thank my husband for nudging me to write this book while constantly questioning the boundaries that I draw for myself. My daughter, Guji, meanwhile, has a penchant for shifting my goalpost several steps away as soon as I am about to reach it. I must thank her for increasing my resilience to failure. I am deeply grateful to my father for cultivating my logical and critical thinking, to my mother for fostering my creativity and empathy, and to my sisters for iterant doses of inspirational conversations.

This text has been coming to life for several years, and it is the patience of the publishing house, Springer, that has brought it to the light of day. The team of professionals at Springer has indulged me with umpteen reminders of work to be done and deadlines to be met. If not for the patient persistence of the Springer team, this project may as well have been shelved.

Several chapters in this book are adapted from the author's previously published works. Chapter 2 is revised from the article "Shape of the Design Worldview: Does Language Inform the Design Sense?" published in Design Issues (2024). Chapter 3 draws upon the article "A Glimpse Inside the Banarasi Mind: What Do Linguistic Categories Reveal About Banarasi Toys?" originally published in Craft Research (2014), with modifications for clarity and context. Chapter 4 incorporates material from the author's book chapter "Dincs as Worldviews: Things that Communicate a Mind," published in Toys and Communication (Palgrave Macmillan UK, 2017), and extends the discussion to include the conventions, canons, and beliefs associated with a broader range of Bāṇārasī Khilōnās. Chapter 5 is adapted from the article "Inexplicit Learning: Transferring Knowledge through Visual and Emulative Practices," originally published in Techne Serien—Forskning i slöjdpedagogik och slöjdvetenskap (2017), and has been revised for clarity and contextual relevance. Additionally, Chap. 6 draws upon material from the author's earlier publication "Artisanship of risk: the uncertainty of traditional design" in *Journal of Vocational Education and Training* (2025). The current version of all previously published papers have been revised for clarity and contextualization. All adaptations have been made with permissions from the publishers, and the chapters have been updated to suit the thematic framework of this volume. I am grateful to all the publishing houses for granting permission to reprint these papers.

Kanpur, India Koumudi Patil

References

Escobar A (2011) Sustainability: design for the pluriverse. Development 54:137–140
Gropius W (1919) Bauhaus Manifesto and programme. The Administration of the State Bauhaus at Weimar, Weimar
Latour, B (2008) A cautious prometheus?: a few steps toward a philosophy of design (with special attention to Peter Sloterdijk). In: Networks of design, annual international conference of the design history society. University College Falmouth, Cornwall, pp 2–10
Mohsini M (2010) Becoming an 'Asli Karigar': the production of authenticity among old Delhi's Muslim Artisans. University of London, School of Oriental and African Studies (United Kingdom)
Mohsini M (2020) Crafting Muslim Artisans: agency and exclusion in India's urban crafts communities. In: Critical craft. Routledge, pp 239–258
Schön D (1983) The reflective practitioner, how professionals think in action. Basic Books
Subramanyan KG (1987) The living tradition: perspectives on modern Indian art. Seagull Books
Teiwes H (1996) Hopi basket weaving: artistry in natural fibers. University of Arizona Press
Wilkinson-Weber CM, DeNicola AO (eds) (2020) Critical craft: technology, globalization, and capitalism. Routledge
Wood G (1985) Labels: a shadow across reality: an introductory note. Dev Change 16(3):343–345

Contents

About the Author

Dr. Koumudi Patil is currently an associate professor in the Department of Humanities and Social Sciences and also participates in the Department of Design at the Indian Institute of Technology (IIT) Kanpur, India. Her research expertise lies in the design and creative processes of the creative industry within India's developmental framework. Through her work, she has spearheaded numerous research initiatives aligned with the Sustainable Development Goals (SDGs), collaborating with the Government of India, industry leaders, and grassroots communities to foster impactful change.

List of Figures

Chapter 3

Chapter 4

Chapter 5

Chapter 6

Chapter 1
Toying with Perspectives: Listening to the Views of Cultural Designers

1 Perspectives on Cultural Design Practice

In 1964, two Australian mineralogists—I. J. Bear and R. G. Thomas—published a paper on the fragrance of the parched earth after the first rains. The paper, titled "Nature of Argillaceous Odour," coined the term *petrichor* (a metonym) from the Greek words *petra* (rock) and *ichor* (the blood of the gods). The mineralogists attempted to explain the formation of petrichor from spontaneous catalysis in controlled atmospheres or from the adsorption of compounds of petrichor. The paper was published in *Nature*—a highly acclaimed international journal that has disseminated the finest peer-reviewed research in science and technology since 1869.

The paper by Bear and Thomas discussed a small perfumery industry based near Kannauj, in Uttar Pradesh, India, as a production center of petrichor. In this cottage industry, the chemists are often uneducated and work with limited and frugal tools and technologies. This particular *ittār* or perfume is called *mittī ittār* or *earth perfume* by them. For about 400 years, the chemists of this little town have used small sun-dried clay bricks for steam distillation in a *deg* or a copper vessel. The vapours containing the odour are collected in a cooling tank (*bhapka*) and absorbed in a container filled with sandalwood oil. The process of distillation is followed with scientific precision, in laboratory-like conditions, without any technical instruments of measurement, and the process yields a fragrance that is unique enough to be mentioned in *Nature*. The process itself and the fragrance produced are indeed remarkable.

Despite rudimentary technology, skills limited to particular materials and tools alone, low literacy, and an absence of foreplanning on the drawing board, the design thinking process followed in Kannauj has produced an astonishing variation in perfumes (Gulāb, Keorā, Hīnā, Champā, Bakul, Pārijātak, Chamelī, Motiyā, Gendā, Rātrānī, Kadam to name a few) that not only fit the needs of the community but also that of the urban, modernized market—largely unfamiliar to the perfume makers. Emperor Jahangir and his wife Mumtāz Mahal were great connoisseurs of the Kannauj perfumes. Today, large firms such as Forest Essentials, Zighrana,

K. Patil, *Toying with the Mind*, Design Science and Innovation,
https://doi.org/10.1007/978-981-96-9227-9_1

and Perfumery brands market these products, and several European perfumers use their olfactory compositions. Despite the contemporary relevance of practices such as those used at Kannauj, and despite highly skilled practitioners, sustainable production cycles, and astute methods of problem solving, cultural design practice[1] continues to be approached from an anthropological, ethnographic, or economic perspective in which cultural designers—also known as craftspeople, vernacular designers, indigenous artists, unselfconscious designers, *shilpī*, and so on—are often regarded as a historical vestige, a specimen for observation, or as members of an ailing enterprise that requires rehabilitation. The cultural design practice is often viewed as comprised of workers from lower-class communities dependent on locally sourced materials; in contrast, the cognitive endeavours of designers and artists are often seen as the sole realm befitting intellectual prowess.

The peer-reviewed discourse on cultural design communities often mentions that their inherent tendencies toward self-organization and self-adjustment may act as constraints on creativity. Alexander (1964, 37–48) acknowledges that cultural practices tend to incorporate changes only when driven by strong compulsions, and in such cases, the need for significant creative effort is often bypassed. When confronted with complex decisions, these practices struggle to generate entirely new forms. Echoing this sentiment, Jones (1970, 19) observes that without the aid of a drawing board, cultural practitioners engage in a continuous process of minor variations and countervariations through trial and error. This method of iterative adjustments, aligning one part with another, inherently limits their work to small-scale, incremental changes. Consequently, innovation tends to remain anchored in precedents, even in scenarios that demand a comprehensive reorganization of form.

In contrast to modern design thinking, cultural design thinking is characterized by a certain inherent fixity. This is due to the influence of myth, tradition, and taboo, which collectively resist deliberate change (Alexander 1964, 48). From a contemporary perspective, a design approach rooted in tradition and habit may seem inadequate for creative problem solving and form generation. This perception likely arises because modern designers often prioritize change or the pursuit of new forms as the foundation of their design endeavours. In contrast, cultural designers begin with the unchanging elements and incorporate change as necessary (Hubka 1979, 28). Athavankar (1989, 104) concurs, noting that the primary intention behind most cultural designs is to conform to accepted norms rather than to exhibit individual uniqueness. Consequently, while the drive for novelty propels the innovation industry, adherence

[1] Please refer to the discussion in the Preface under the section—"What Should It Be Called?: The Problem of Labelling" where I have attempted to define this term. To retain a semantic respect for the craft profession, I choose to use the term *cultural design practice* in this text. This term acknowledges the ability to design, cogitate, and innovate in the craftspeople, albeit differently from their formally schooled or industrial cousins. The culture here is not a fixed, rigid practice but what Subramanyam (1987) would call "living." Living practices are continuities of creative practice that evolve and change in response to context. The arguments made in this text will assert the reflexivity of cultural design practice and reposition handmade practitioners as designers of abilities to solve problems responsibly and innovatively in ways that are equal to yet different from the abilities of their modern, city-dwelling cousins.

to typicality remains a significant criterion for acceptance within cultural design communities.

Evaluating the discipline of cultural design against the standards of modern design is a historical fallacy and is particularly evident in the post-industrial period's efforts to protect handmade designs from the perceived detriments of mass production. Modern design, an outgrowth of the Industrial Revolution, developed to meet the demands of a machine-dominant, high-speed context through processes inherently misaligned with those of cultural design. Therefore, expecting cultural design to conform to the criteria of modern design—such as precision, constant innovation, standardization, mass production, minimalism, cultural neutrality, artificiality, explicit processes, and theoretical frameworks—is misguided.

Camnitzer's vehement opposition to such colonization, particularly in the field of printmaking, aptly captures this inequitable dynamic. He states, "When I refer to colony, I mean it quite literally: as a territory taken over by another, where identity is maimed and slowly forgotten, values are shifted and the will for independence becomes ritualized into an increasingly empty and hopeless vow" (Camintzer 2009, 104). This analogy underscores the fundamental incommensurability between the situated practice of cultural design and the framework of modern design to revive and protect the former.

It is imperative to develop characteristics intrinsic to cultural design to liberate it from the burden of emulating modern design. It is time to recognize and appreciate the distinct methodologies, actions, and philosophies that underpin cultural design.

In recent times, cultural design, along with the designers and their distinctive problem-solving methodologies, has reentered contemporary discourse as an alternative approach to cognition and visualization. Cultural designers are distinguished not merely by their practiced skill but by their remarkable capacity to improvise decisions amidst the unpredictable circumstances often encountered on the workshop floor. Much like modern design thinking as explored by Cross (1982, 2006, 2011), the interplay of problem finding and problem solving is central to cultural design thinking. However, in cultural design, these processes also run concurrently with the act of creation itself. While cultural design frequently emulates previous designs, it equally engages in spontaneous problem solving, often devoid of prior planning. When unforeseen challenges arise, the cultural designer must adapt in real time (Kritzer 2007, 328). Consequently, problem finding and problem solving are intimately linked within culture-based artisanship, where errors are frequently transformed into creative opportunities (Sennett 2008, 9). This dynamic melds rigid traditions with immediate action and fosters a self-regulating process. Such a process embodies an intrinsic equilibrium that guides even the most haphazard alterations into cohesive forms (Alexander 1964, 52). In this context, the cultural object influences thought, yet thought is simultaneously manifested through the hands shaping the object. Thus, cultural design thinking, akin to any mental process that materializes into an artefact, holds significant relevance within the discourse of designerly[2] knowledge. Cultural designers are, in essence, the early architects of the artificial world (Simon 1969), though their modes of knowing, thinking, and acting may diverge. Hence, it is plausible to investigate the methodologies underpinning cultural design thinking in a manner similar to inquiries into modern design ways of thinking.

This text endeavors to reposition cultural design within the contemporary design framework, asserting it as a unique, contextual worldview. It presents cultural design as a method for solving problems within local, rather than global, boundaries; as a pedagogical practice for transferring tacit knowledge into explicit applications; and as an innovation ecosystem that prioritizes cultural acceptability over disruption. In doing so, the text will navigate between two divergent frameworks of design.

The first framework posits that cultural design is an iterative practice that has evolved an inherent equilibrium that guides even the most aimless changes into harmonious forms. In contrast, the second framework contends that cultural design is a creatively restrictive practice resistant to change. Numerous examples of varying skill levels and design innovation within the cultural design communities of India substantiate both perspectives.

During numerous visits to the cultural design communities in Uttar Pradesh, India, for a government of India project, I observed veteran, ageing cultural designers adorned with many state and national awards proudly showcasing their skills and lineage. Their faces radiated with pride as they spoke of the immense time and effort invested in their cultural practices. For instance, the late Muhammad Tughlaq of Morādābād dedicated eleven years to creating a jewelry box that ultimately won him the National Award (Fig. 1). This simple rectangular silver box is intricately engraved with a scene depicting a tiger hunting its prey—a deer—in a dense forest. The *nakāshī*, or engraving of the foliage and animals, maintains consistent size, shape, and proportions throughout the box. Tughlaq revealed that he simply picked up his tools, and the image flowed seamlessly into his seasoned hands. This unique artefact remains without replicas. Tughlaq priced this meticulously carved box at 11 lakhs (approximately $20,668.89 US as of February 1, 2013), which amounts to one lakh for each year of his work.

Jones (1970, 15) marveled at the beauty and complexity of such artisanship, which could easily be mistaken for the work of a highly skilled designer even though the piece was created without the support of managers, salespersons, production engineers, and the myriad other specialists upon whom modern design depends. How can an illiterate cultural designer, equipped with simple tools, without a drawing board for trial and error, and with no visible coding or patterns from which complex forms can be derived, achieve such coherence in an artefact?

Tughlaq and numerous scholars from previous generations attribute this tacit expertise to a divine gift bestowed by the gods upon the chosen few. Kramrisch (1958, 226) elucidates that the traditions of Indian designers are the methods and practices

[2] Designerly thinking is a practice-based approach to solving problems, making sense of things, and developing new knowledge based on abductive processes (Buchanan 1992). The term was coined by Cross (2006).

Fig. 1 *Nakāshī* on a jewellery box designed by Late Muhammad Tughlaq in Uttar Pradesh, India. *Image source* Author

through which they manifest their professional activity and apply their understanding of the Principle. This Principle is the source and origin of their vocation, known as *Brahmā* or *Viśvakarma*, and embodies the totality of creative consciousness. Indian cultural designers perceive their art not as their own nor as the accumulated skill of generations, but as originating in the divine skill of Lord *Viśvakarma* himself. Therefore, the terms *artist*, *artisan*, *designer*, and *craftsperson* are not apt translations of *Shilpin* (a literal translation of *craftsperson* in Hindi); for in India, the arts, crafts, and design process partake in rites whose technical performance holds magical power.

The materials and tools of cultural design are considered sacred and are believed to house particular powers (refer to Transcription 1). *Āyudha Pūjā* or *Astra Pūjā* is a day when Indian cultural designers worship their tools with incense, flowers, and unhusked rice—a ritual that acknowledges and accepts the otherworldly powers embodied in these inanimate objects through which creativity is expressed.

"We are *Pañcāl Brāhmins*. Our ancestors were the Pāṇḍavas. Our lineage is connected with the *Pañcāl Nāresh*—the father of Draupadī. We pray to Lord *Viśvakarma* and to our tools. *Viśvakarma* was the incarnation of Lord Viṣṇu himself. He had five sons—Manu, Māyā, Tvaṣṭṛ, Śilpī, Viśvajña. They became the *Lohkār*—ironsmith; *Kāṣṭhakār*—carpenter; *Śilpakār*—cobbler; *Swarnakār*—goldsmith; and *Kaserā*—utensil maker. It is according to their work that *Viśvakarmas* are divided. ... It is beyond my imagination how long ago we go back to... Rāvaṇa's Laṅkā [which] was also made by *Viśvakarma*."

Transcription 1: Shyam Sundar *Viśvakarma*, a *Bāṇārasī* toy designer from Varanasi, India, speaks of his lineage to Lord *Viśvakarma*.

Even the raw materials are animated with life and energy. *Bṛhat-saṃhitā* (composed by Varāhamihira in the sixth century in the Sanskrit language) urges the carpenters, architects, and sculptors working in wood to seek forgiveness from the tree that they need to cut for designing the artefacts. Such respect for the material world around us takes the discourse of sustainability to a level higher than a mere transactional approach for survival.

अर्चार्थममुकस्य त्वं देवस्य परिकल्पितः।

नमस्ते वृक्ष पूजेयं विधिवत् सम्प्रगृह्यताम्।।10।।

यानीह भूतानि वसन्ति तानि बलिं गृहीत्वा विधिवत्प्रयुक्तम्।

अन्यत्र वासं परिकल्पयन्तु क्षमन्तु तान्यद्य नमोऽस्तु तेभ्यः।।11।

Śloka 10–11.—Oh Tree, you have been selected for the worship of the particular deity. Salutation to you! I worship you in accordance with the rules, kindly accept it. May all those beings that dwell in this tree fix their residence elsewhere after receiving the offerings made according to rules. May they forgive us now; We bow to them.

Since cultural design skills and processes are perceived as stemming from a divine source, masters like Tughlaq find themselves unable to explicitly codify them. Without any explicit transfer of generational practice, the tacit knowledge of these masters perishes along with them. Sennett (2008, 77–78) recounts the tale of Stradivari, the renowned eighteenth-century designer of stringed instruments. Upon his death, he left his workshop to two apprentices, but the workshop eventually faltered. Stradivari had neither taught nor could he teach the apprentices how to be a genius. The master's originality, individuality, and distinctiveness in producing complex objects remained personal secrets in the absence of means to transfer them explicitly.

Nonetheless, the passing of generational skills from master to apprentice can result in beautifully evolved artefacts. For example, the Hopi basket weavers, Native Americans from Arizona, adapt their designs to meet the needs and desires of tourists and collectors. Teiwes (1996, 159) was granted access by the Hopi tribe and their Cultural Preservation Office to observe their community and rituals—information rarely available to outsiders. Teiwes records that the cultural design of "deep" baskets had inward-curved openings to protect stored food better. However, this design was impractical for American urban customers, who preferred wider openings for other uses. Consequently, Hopi basket weavers adapted their cultural knowledge to create large, "deep" baskets with straight sides and wide openings, which allowed the baskets to be used as wastebaskets with lids. Such evolutionary instances of generational knowledge in cultural epistemes have produced enduring forms.

Unlike Tughlaq, Stradivari, and other lesser-known or unknown masters who failed to transfer their tacit knowledge to apprentices, George Sturt, a rare literate cultural designer, published a book in 1923 on the design of the wagon maker. It is from this book and field observations that Jones was able to discuss the master's role in relation to an artefact. Jones (1970, 16–24) explained that in the absence of a drawing board, cultural designers typically make a series of minor variations and counter variations, continuously adjusting one part to another, with the result that no two products are alike. From this process of design evolution, Jones derived several insights into cultural design thinking:

1. Cultural designers do not know the reasons for what is done, but they know how to do it.
2. The form of a cultural design is modified by countless failures and successes in the process of trial and error that spans centuries in an ongoing effort to correct mistakes or meet new demands. This slow search for the "invisible lines" of a good design, in the end, can produce an astonishingly well-balanced result and a close fit to the needs of the user.
3. Cultural design evolution can also produce discordant features if faced with sudden changes. This is an inherent weakness of changing only one thing at a time and of relying on precedents even in contexts that require a complete reorganization of the form as a whole.
4. Fragmentary information is stored as patterns and as exact memories. It is learned during apprenticeship as the apprentice learns the actions needed to re-create the cultural shape of the product.

In this design process, the division of tasks among cultural designers is only feasible when critical dimensions are predetermined or standardized (Jones 1970, 22). This limits the work of cultural designers to smaller dimensions and incremental changes. Changes in cultural design are slow, improvisatory, contingent on trial and error, and heavily reliant on precedents, even in contexts demanding a complete reorganization of form. From a modern perspective, a design strategy based on tradition and habit may appear to be an inadequate method for creative problem solving and form generation. Alexander (1964) acknowledges that when confronted with complicated choices, the cultural designer struggles to invent forms from scratch. While this is largely true, it does not account for the cultural design processes involved in creating houses and wagons, which are large artefacts that require advance planning and judgment.

In the context of cultural artisanship, it's essential to recognize that not all cultural designers are driven by the intrinsic satisfaction of meticulous artisanship. Many are motivated by the practical need to sustain themselves in a contemporary marketplace. Unlike legendary figures such as Tughlaq and Stradivari, who are celebrated for their extraordinary and mysterious abilities, Ramdhenu Viśvakarma, a cultural metal designer from the modest town of *Mahobā* in Uttar Pradesh, India, reproduces counterfeit figurines based on designs passed down from his ancestors. These inexpensive, imitation figurines encapsulate cultural elements intended for tourists, who eagerly purchase them. Ramdhenu, in contrast to Tughlaq, neither takes pride in his

(a) *Nakāshī* on a jewellery box. (b) Fake casts of Mahobā brassware.

Fig. 2 *Morādābād* and *Mahobā* brassware observed in local *Karkhanās* during a field visit in 2008 in Uttar Pradesh, India. *Image source* Author. Both images were retrieved in 2014 from the blog created by Chris Dame, Manjari Sheel, and Soujanyaa Boruah to document the process of making these toys. The blog is called "Little Known Heroes", and its link is http://blog.littleknownheroes.com/2011/06/o-holy-artisan-where-art-thou/

inherited cultural practice nor strives to learn new skills that could provide alternative means of livelihood (Fig. 2b). He is satisfied with replicating a skill he memorized during his childhood—a routine that has become his means of subsistence.

It is a bitter irony, and indeed symptomatic of broader trends, that the burgeoning interest in cultural design coincides with the moment when many studio-based arts and generational designs are classified as "Endangered Subjects" (Ferris 2009a, b). The decline of cultural artisanship not only threatens livelihoods but also erodes communal knowledge, intergenerational transfer of skills, and traditions. In cultural design communities, the transmission of knowledge and skills occurs informally within the workshop from seasoned masters to young apprentices. The disruption of this generational transfer due to declining consumption results in a profound and irreplaceable loss of production and consumption knowledge (Balaram 1998; Jaitly 2012).

Faced with these challenges, many cultural designers opt for the relative security of daily wages under the Mahatma Gandhi National Rural Employment Guarantee Act 2005 (MGNREGA) and engage in unskilled labor such as digging or road cleaning. This choice often seems preferable to the grueling conditions of generational artisanship—enduring extreme heat, straining over intricate embroidery or loom work, or risking silicosis from stone carving. Moreover, the compensation for these generational skills no longer reflects the effort and creativity invested. It is, undeniably, an unfair trade.

Given these circumstances, it is not surprising that many cultural design communities in India have either adapted their skills to align with Western tastes, thereby losing their cultural status, or are struggling to survive due to declining demand in the modern urban market. This raises critical questions: Is the decline of cultural artisanship, exemplified by figures like Tughlaq, an anomaly or merely part of a broader spectrum wherein cultural designers like Ramdhenu exist at the lower end? The balance between skill and technology has been disrupted; technology has triumphed, and the general skill level of cultural designers has diminished (Ferris 2009a, ii). Has the economic dominance of the machine monopolized the market, or are cultural designs floundering because they are ill-suited for the modern urban context?

These questions, along with the contradictions in the literature and the urgent need to address the rapid decline of cultural design knowledge, have driven the writing of this text. As a result, the text delves into the minds of cultural designers in an attempt to understand the thoughts and motivations that propel them into action, adaptation, and, at times, failure.

2 Packaging Culture for a Modern Market

The *Bāṇārasī* community of wooden toy designers[3] on which this book is based reflects a similar variation in cultural artisanship and thinking among its members. It is one of the rare surviving cultural design communities that boasts of both a domestic as well as an export market despite the onslaught of mass-produced, cheap, Chinese plastic equivalents. Interestingly, the community has preserved its traditions and also changed with the times without any external assistance from the government or non-governmental organizations (NGOs). Despite the odds, the community's resilience raised my curiosity about its means or strategy of survival in a globalized scenario.

The *Bāṇārasī* community of wooden toy designers is nestled on the banks of the holy river *Gaṅgā* at Varanasi—also known as *Kāśī* or *Bāṇāras* (in this text, I will refer to the place as *Bāṇāras* and its people and objects as *Bāṇārasī*, as observed in the local dialect)—in the northern state of Uttar Pradesh, India. The community largely consists of about 1000 *Karkhanās* (homes that also serve as workshops) that are traditionally run by *Viśvakarma* and the *Prajāpatis*. (These are Hindu castes associated with creation and artisanship. Recently, other castes, such as *Sharmās* and *Mauryas*, have also joined the community.) There are numerous space and place mythologies about when, how, and with whom the wooden toy community of *Bāṇāras* began. One account suggests that the *Bāṇārasī* community of cultural designers who work in clay, primarily consisting of *Prajāpatis*, was introduced to wood turning from Udaipur, Rajasthan, at the behest of the then ruler of *Kāśī* about 400 years ago (Ranjan and Ranjan 2007; Srivastav 1984). Wood-turned toys have been designed in *Bāṇāras*

[3] Bāṇāras is renowned for not only its toys but also *sarees*, *gulabi meenakari*, bead making, clay toy making, *zardosi*, *repousee*, silver work, as well as a small number of ivory carvers and boat makers. The producers and customers of all these cultural design artefacts belong to different communities and sometimes different religion too, each with its own unique set of traditions taboos and rules. These cultural design communities rarely interact with each other and are self-sufficient. The influence, if any, is minimal. Some have been studied by Ranjan and Ranjan (2007, 158–163) and Jaitly (2012, 124–139). In order to make a rigorous enquiry, this text focuses on only the wooden toy makers. References to other communities are made only when they influence the practice of toy making.

to this date. A different version, orally narrated by the Agarwal family, one of the first family of entrepreneurs of wooden toys in *Bāṇāras*, recounts that the ban on ivory in the 1900s forced the ivory carvers to shift to wood. The skilled ivory carvers adapted their artisanship to hand-carved figurines in wood called *Gadhwā kām*. *Gadhwā* indicates digging or sculpting. In a third oral account, the clay toy makers also claim woodwork dating back to about 100 years ago. The senior toy designers—*Gopal Prajāpati* and *Ramprasad Prajāpati*—recall the temptation of shifting to wood in the 1920s following the replacement of clay by Plaster of Paris. Increasing social pressure to abandon the use of furnaces (due to the rising pollution in the city) and the ease and profit of woodwork also catalysed the shift to wood. In a longitudinal ethnographic study conducted over ten years from 2008 until 2018 in the *Bāṇārasī* community of toy designers, I visited several *Karkhanās* (or studio workshops and local shops). I met toy designers from several communities, local customers, and experts endowed with social and moral authority in the community—experts such as *Pandits* (priests) and relatives of the royal family of *Kāśī* (the ancient name of the city of Varanasi) who were patrons of the design community. This is a closely knit community wherein the cultural designers are related to each other via caste, master–apprentice relationships, or direct family kinship, and to the shopkeepers via inter-generational transactions. The shops at *Vishwanath gali*[4] are inherited by local entrepreneurs from their forefathers, who also sold toys designed by this community. For instance, the owner of Agarwal Toy Emporium describes how his business was the first to receive an order to produce Russian nested dolls in the 1920s, an endeavour for which wood-turners were hired. The shopkeepers share a close bond with the toy designers, who assist them financially and socially in troubled times.

The local customers are also strongly dependent on the cultural toy market for fulfilling their needs of play as well as rituals; however, their relationship is purely transactional. Several festivals cannot be celebrated in *Bāṇāras* without these wooden toys. I will discuss the role of toys in the *Bāṇārasī* community in more detail in Chap. 2. However, for the purpose of the present discussion, it suffices to say that the toy designers, toy sellers, and toy patrons together form a community of practice. This naturally formed community of practice[5] shares a common passion for play and toys. The toy makers constantly learn and modify their knowledge to meet the needs of the patrons, and the shopkeepers and patrons interact regularly with the designers for transactional exchanges and for sharing constant feedback. Together, they all practice the know-use and the know-how of toy designing.

---

[4] Street.

[5] The concept of communities of practices was coined by cognitive anthropologist Jean Lave and educational theorist Etienne Wenger in 1991 in their seminal work titled *Situated Learning—Legitimate Peripheral Participation*. Later, Wenger also significantly elaborated on this concept in her 1999 book *Communities of Practice*. It is well accepted that communities of practice are not new phenomena: this type of learning has always existed because people practice, learn, and share their experiences with each other. C. S. Peirce's concept of the "community of inquiry" (Shields 2003) and John Dewey's principle of learning through occupation (Wallace 2007) precede the work of Lave and Wenger. From this perspective, Cultural design communities are one of the oldest learning communities in the world.

However, the increasing replacement of the traditionally familiar patron of the toys from within the community with an unknown customer outside the community makes the problem of design decision making complex. The customer segment from outside the community, consisting either of exporters or the urban Indian customer, is not only unfamiliar with the cultural design practice but also demands a pace and kind of change that is unfit for cultural production. It is incorrect to assume that this is the result of a consumerist culture borne out of a post-industrial society. Such demands and pressures have always been a part of cultural design practices. In 1909, Edwin Greaves, in his book *Kashi the City Illustrious Or Benares*, describes the poor imitations of Western designs in Bāṇārasī toys. Such unskilled designs are called *Challu kām* or mediocre work in the local dialect today.

> A third industry (in *Bāṇāras*) which employs a few of the people is lacquered wood-work. The articles are made almost exclusively of toys and various small boxes, especially the 'nests' of boxes, a large number of tiny boxes that fit one within the other. The lacquering is still on the lathe. Practically all these toys are turned. At present, however, a rage is setting in for painting flowers, etc., on these toys, to the no small detriment of their artistic worth, for, as a rule, the flowers and other designs painted on them are but crude and tawdry work, and indicate the retrograde movement observable in so many indigenous industries, where the purely Indian type of work is displaced in favour of feeble imitations of poor specimens of western productions. (Greaves 1909, 25–26)

A contemporary example is the case of Little Known Heroes, a 2012 project undertaken by three formally educated modern designers: Chris Dame from Quicksand Design Studio in the United States, and Manjari Sheel and Soujanyaa Boruah from New Delhi. The designers asked *Bāṇārasī toy* designers to make wooden figurines based on 2D designs created on software. These characters were drawn from Indian mythology and had no precedent in the *Bāṇārasī* visual culture. The shapes, as well as the styles and motifs were completely unfamiliar to the toy designers. Moreover, the paper designs included only the front profile of the figurines. Despite the lack of information, the toy designers visualized the entire design in 3D and conceptualized and made a back profile in the same visual language as the front profile. In yet another order in the same consignment, Chris Dame, Manjari Sheel, and Soujanyaa Boruah gave a *plain*, lathe-turned figurine to a toy designer and asked him to carve out details as he pleased, with no further instructions on motifs, size, kind of carving, and so on. Nevertheless, working with an unfamiliar shape, the toy designer carved a beautiful, cohesive design without any preconceptions on a sketch pad or rehearsal on a low-fidelity prototype. Refer to Fig. 3.

This phenomenon of unfamiliar consumption patterns is not unique to *Bāṇāras*. In 2008, the proprietor of *Sainik Farms*—an export unit based in Morādābād,[6] India (a cultural design community that makes brass artefacts)—bought the skull of a bison from a European provider. The proprietor observed that bison skulls had a lucrative

[6] *Morādābād* is a cultural design industry in the state of Uttar Pradesh, India, that came into prominence under Mughal patronage in the late seventeenth century. The design involves traditional techniques such as casting, engraving, and polishing and often results in exquisite patterns and designs. The brassware industry is predominantly Muslim and is divided into an assembly line system of skill sets.

(a) (b)

Fig. 3 **a** Modern designs of Indian deities made by Chris Dame, Manjari Sheel, and Soujanyaa Boruah. **b** Interpretation of the modern designs by *Bāṇārasī* toy designers. *Image source* Retrieved from http://blog.littleknownheroes.com/2011/06/o-holy-artisan-where-art-thou/ in 2014, created by Chris Dame, Manjari Sheel, and Soujanyaa Boruah

market among the bohemian sector of European youth. Therefore, for the order placed by the proprietor, hundreds of bison skulls were cast in aluminum (Fig. 4) by the *Morādābād* brassware designers—a far cry from the abstract or floral *Nakāshī* of the brass vases for which masters like *Muḥammad Tughlaq* are renowned. These bison skulls are not seen in the local *Morādābād* market because Islam forbids designers from producing graven images. Therefore, the skulls were not sold in the local market to *Morādābādī* patrons but were nevertheless produced for a foreign customer that sought profit.

During numerous field visits to *Bāṇāras*, it has been striking to observe that a single cultural designer does not produce a toy independently. Instead, like an assembly line system, the process involves the skills of multiple participating members. Though the toy designers share a common language—*Bāṇārasī*—they lack explicit methods for sharing or exchanging information among group members who work in physically separated workshops. Similar to a relay race, each succeeding toy designer continues designing the toy based on the work of the previous team member. Despite the involvement of multiple decision makers in the design process, the toys maintain a remarkable coherence.

These generational design skills, although they incorporate a division of labor, are rooted in meaningful decision-making abilities that are contextually adaptable to new concepts and cannot be replaced on short notice. They also entail a long learning trajectory. In contrast, the Fordian division of labor developed by auto maker Henry Ford emphasizes economy of production, optimization of skilled labor, and segregation of the skill of thinking from the labor of making. By allocating skilled tasks with the freedom to improvise within the shared worldview, the *Bāṇārasī* toy design practice has utilized the assembly line system as a tool for change and innovation while meeting the large-scale production demands of export clients. Consequently, every toy designer is capable of introducing changes or building upon previous alterations within the boundaries of the worldview.

Fig. 4 Bison skulls cast in aluminum in *Morādābād*, India. *Image source* Author

In the absence of a single leader or centralized planning, *Bāṇārasī* toy designers produce, share, and evolve their generational knowledge of designs. How does such leaderless thinking and production emerge to conceive designs? It is even more intriguing to understand how multiple members separately conceptualize different parts of a wooden toy without a leader, a centralized vision, or explicit coordination and yet achieve coherence. This challenges the "myth of the centralized mindset." Resnick (1997) argues that the centralized mindset assumes a task can be solved most efficiently by a single entity governing the task flow and decisions continuously. Such assumptions are commonly found in mainstream design practices, which are often coordinated by a leader who facilitates, organizes, and envisions the solution to a problem either independently or by assigning pre-decided tasks to other team members. While explicit centralized planning has its advantages, it is equally fascinating to observe that leaderless groups of *Bāṇārasī* toy designers are creating toys not only to preserve their culture but also to innovate. The question remains: how?

The phenomenon of leaderless production in the absence of or with limited knowledge about an unfamiliar customer is itself a subject worth investigating. Such instances challenge the notion that coherence alone suffices in the design process. *Bāṇārasī* toy designers, for instance, produce Christmas decorations—crosses, stars, figurines of Jesus, and reindeer—in large quantities for the European and American markets every year from September to November, despite lacking local patronage for these items. This requires the designers to possess a strong visualization capability to make decisions that cater to unfamiliar customer needs. Additionally, they must operate with a well-rehearsed set of implicitly shared notions to ensure multiple designers can make mutually coherent decisions without explicit planning and coordination.

While the Christmas designs produced in *Bāṇāras* for export to Europe and America are coherent, they starkly contrast with the *Bāṇārasī* identity of traditionally made toys. In fulfilling such client orders, *Bāṇārasī* producers are not creating *Bāṇārasī* toys but merely replicating American or European artefacts. This loss of *Bāṇārasī* identity is intrinsically linked to the erosion of cultural design itself. Is this new consumption pattern undermining the identity of *Bāṇārasī* cultural design? The dominance of mainstream design sensibilities and market demands raises critical questions about the survival of *Bāṇārasī* cultural design, its designers, and their unique way of thinking as they balance earning a livelihood with preserving their identity.

Modern consumption patterns are neither familiar to nor in alignment with the worldview of the *Bāṇārasī* community. Yet, in certain instances, *Bāṇārasī* designers have designed toys despite lacking complete information about the preferences of their unfamiliar customers. The challenge is exacerbated by the practice of multiple producers independently conceptualizing different parts of the toy. For cultural design, toy designers rely on their collective worldview to make informed conjectures in the absence of a shared sketch or plan. However, a non-*Bāṇārasī* consumption pattern does not fit within the community's worldview, leading to a lack of a coherent idea to unify the final design within an assembly line system of production.

The good fit between form and context that Christopher (1964) advocated is more difficult to achieve for an unfamiliar market—an unfamiliar context. This prompts us to question the widely held belief that cultural design communities resist willful change. If that were true, how can the *Bāṇārasī* community of toy designers design toys catering to unfamiliar modern preferences of customers?

3 How to Toy Around with This Text

As the *Bāṇārasī* cultural designers confront the unprecedented challenge of designing for an unfamiliar and unknown customer, and do so without a central vision, leader, or explicit coordination, it becomes imperative to examine how these toy designers are adapting their thinking to this new scenario. This is especially relevant given the stronghold of traditions and resistance to willful changes within their community. The appropriateness of the design decisions shaping the nature and extent of changes in the toys is crucial for the survival of cultural design, cultural artisanship, and cultural design thinking in this new context.

Therefore, in this text I discuss *Bāṇārasī* toys from a designer's perspective; I aim to elucidate the design decisions involved in conceiving and making a design by the cultural designers themselves. I delve into the minds of cultural designers to understand the notions that spur them into action and change; I present reasons for their success or failure in creating an artefact. Hence, I discuss cultural design not from an anthropological, political, or philosophical perspective as a way of life that waned with the advent of industrial society, but as a contemporary discourse of thinking and making a design.

With this line of thought, all arguments and field data in this study related to consumption, social conventions, beliefs, rituals, and myths held by the *Bāṇārasī* toy designers are used solely to support the decisions of a cultural designer in thinking about and making a cultural design. Given that the worldview of every cultural design community is unique to its own context, the scope of this study and its conclusions are limited to the *Bāṇārasī* toy designers alone.

The book is organized into nine chapters, beginning with the introduction presented in this chapter. Chapter 2 delves into the framework of cultural design thinking through the lens of a collective disposition or community-held worldview. I argue that the worldview of a community both shapes and is shaped by design. In the case of the *Bāṇārasī* toy designers, design decisions are not made by an individual but by a community of cultural designers. Therefore, this text aims to elucidate the thinking of multiple minds involved in the making of a *Bāṇārasī* toy. Consensus in decision making among multiple minds, in the absence of a common plan, implies a shared set of implicit underlying principles or perceptions within the community that informs the explicit practice of making a toy. This community worldview is captured through the *Bāṇārasī* language, which is characterized by an evolving lexicon and a community-held system of categorization.

However, the typical enculturative *Bāṇārasī* toys also display a variety that cannot be explained by changes in lexicon alone. Chapter 3 decodes the community-held categorization of toys from the local lexicon and verifies it through card sorting with the cultural designers. This reveals not only the organizational principles used by the designers but also the distinctiveness of the toys. The relationship among these multiple categories is maintained by the worldview, which determines and limits the acceptability of variations in a toy design within the community.

The rules that determine the relationship between multiple categories are an engaging question to explore. Consequently, Chap. 4 extracts a set of conventions, canons, and beliefs that govern the acceptable combinations of multiple categories within the worldview. At the intersection of these conventions, canons, and beliefs, the *Bāṇārasī* identity is constructed. Chapter 5 is a pedagogical inquiry into the transfer of tacit knowledge between master and apprentice in the absence of explicit coordination or planning. This chapter reconstructs the role of observation and emulation in learning, which preserves the *Bāṇārasī* identity while also facilitating changes in toys that lead to the enhancement of cultural artisanship. Pedagogy concerns not only the transfer of skills and worldviews but also the capability of taking design risks within the community-held categories.

Chapter 6 challenges the myth that innovation is absent in the cultural design sector; it argues that there is indeed a cultural practice of taking design risk that modern systems often underrate. This chapter builds upon the notions of "workmanship of risk" proposed by Pye (1968) and the knowledge spiral of Nonaka and Takeuchi (1995).

Today's cultural design community faces an interesting dilemma. Preservation of the *Bāṇārasī* identity is crucial for cultural designers to maintain their worldview. However, with declining local consumption, toy designers are encouraged to create designs that do not conform to the *Bāṇārasī* identity; this practice can disrupt the worldview. Chapter 7 elucidates several strategies the community has developed to balance their worldview with their livelihood.

Based on the exploration of the *Bāṇārasī* practice of designing, thinking, and consuming toys, Chap. 8 discusses a set of characteristics or signifiers by which we define, examine, intervene, and adjudge the future viability of cultural design, thereby making this polythetic category more inclusive, semantically rich, and possibly more holistic.

Chapter 9 situates cultural design thinking within contemporary discourse in an effort to reframe cultural design as an alternative design practice rather than merely an instrument of archiving a past.

All chapters can be read independently of their preceding or succeeding chapters, thereby allowing the reader to saccade through the chapters rather than peruse the topics in a linear pattern. This flexibility in reading can lead to multiple interpretations of the discussions in this book. I propose to follow this text with a second text that will explore how categorization assists in design decision making, particularly in cultural design thinking. This forthcoming work may tentatively close the circle I have begun to draw.

References

Alexander C (1964) Notes on the synthesis of form. Harvard University Press

Athavankar UA (1989) Categorization… natural language and design. Des Issues 5(2):100–111

Balaram S (1998) Thinking design. National Institute of Design, Ahmedabad

Bear IJ, Thomas RG (1964) Nature of argillaceous odour. Nature 201(4923):993–995

Buchanan R (1992) Wicked problems in design thinking. Des Issues 8(2):5–21

Camnitzer L (2009) Printmaking: a colony of the arts, chap 12. In: Weiss R (ed) On art, artists, Latin America, and other utopias. University of Texas Press, New York, pp 104–111

Cross N (1982) Designerly ways of knowing. Des Stud 3(4):221–227

Cross N (2006) Designerly ways of knowing. Springer, London

Cross N (2011) Design thinking: understanding how designers think and work. Berg, Oxford, New York

Ferris M (2009a) Editor's introduction. Making futures: the crafts in the context of emerging global sustainability agendas. University of Plymouth, Devon, UK, pp i–vi

Ferris N (2009b) Native-lived colonialism: challenging history in the great lakes. University of Arizona Press, Tucson

Greaves E (1909) Kashi the city illustrious, or Benares. Indian Press

Hubka T (1979) Just folks designing: vernacular designers and the generation of form. J Archit Educ 32(3):27–29

Jaitly J (2012) Craft atlas of India. Niyogi Books

Jones JC (1970) Design methods: seeds of human futures. Wiley, New York

Kramrisch S (1958) Traditions of the Indian craftsman. J Am Folklore 71(281):224–230

Kritzer HM (2007) Toward a theorization of craft. Soc Legal Stud 16(3):321–340

Nonaka I, Takeuchi H (1995) The knowledge creating company. Oxford University Press, New York
Pye D (1968) The nature and art of workmanship. Cambridge University Press, Cambridge
Ranjan A, Ranjan MP (2007) Handmade in India: a geographic encyclopaedia of India handicrafts. Council of Handicraft Development Corporation, New Delhi
Resnick M (1997) Turtles, termites, and traffic jams. MIT Press, London
Sennett R (2008) The craftsman. Yale University Press, New Haven
Shields PM (2003) The community of inquiry: classical pragmatism and public administration. Adm Soc 35(5):510–538
Shrivastava V (1984) Pital evam kashta kala udyog. In: Rai K (ed) Kashi ank: Uttar Pradesh. Saral Printing Press, Lucknow, pp 107–110
Simon HA (1969) The sciences of the artificial. MIT Press, Cambridge, MA
Subramanyan KG (1987) The living tradition: perspectives on modern Indian art. Seagull Books
Teiwes H (1996) Hopi basket weaving: artistry in natural fibers. University of Arizona Press
Wallace S (2007) Managing behavior in the lifelong learning sector. Sage

Chapter 2
Language and Shape of the Worldview

1 First Came the Word and then the World

According to Fermat's principle of least time, the path taken by a ray of light to traverse from point A to B is the one that requires the least time to traverse. A fictional character, Dr. Louise Banks[1] (a linguist), extrapolated from this principle that a ray of light has to know its destination before it can choose the fastest possible route among the several available alternatives. She applied this principle in her quest to decipher the language of some Heptapods, or aliens who had landed on Earth. She postulated that the Heptapods do not write a sentence sequentially, one semagram at a time, but simultaneously, and therefore they must know the entire sentence before they begin to write. Furthermore, because the language of the aliens requires simultaneous expression of all thoughts in a sentence at once, the Heptapods' experience of the world must also be teleological, and not causal as it is for humans. This led to her discovery that the Heptapods know the future.

The linguistic analysis of Dr. Banks implied that by deciphering the language of the Heptapods, humans would know how the Heptapods constructed their knowledge, thought, sensed, perceived, and structured their world. Similar to the Heptapods'

[1] The character of Dr. Louise Banks was introduced in the science fiction novella *Story of Your Life* written by Ted Chiang and published in 1998. In 2016, the story was adapted in the movie *Arrival* directed by Denis Villeneuve.

This chapter is adapted, with minor revisions, from the author's article published in 2024 in Design Issues, titled "Shape of the Design Worldview: Does Language Inform the Design Sense?" with appropriate permissions (Patil, Koumudi. "Shape of the Design Worldview: Does Language Inform the Design Sense?", Design Issues 40:3 (Summer 2024), ©2024 Massachusetts Institute of Technology, pp. 88–104.)

A version of this chapter was also published as a part of the proceedings of the Making Futures conference held at Plymouth in 2013, titled "Language and the Shape of the Worldview." The paper has been adapted and revised with permissions to fit the context of this book.

K. Patil, *Toying with the Mind*, Design Science and Innovation,
https://doi.org/10.1007/978-981-96-9227-9_2

conception of time as simultaneous, the Hopis, a Native American tribe, do not conceive of time as linear or sequential. The Hopis do not distinguish between the present, past, and future time of occurrence of an event. Time in the Hopi world is endless and is expressed as such through their language. According to Whorf (1944), in the Hopi language, there are no words, grammatical constructions or expressions that denote time; the language is tenseless. The Sapir-Whorf hypothesis (Hoijer 1954), which is largely based on the examination of this tribe, claims that language shapes, and at times even limits, our perception of the world.

Several studies have supported the theory of linguistic relativity. While studying the language of Guugu Yimithirr, an aboriginal tribe from Australia, the linguist Guy Deutscher (2010) discovered that the language of Guugu Yimithirr describes everything as being geocentrically based on the cardinal directions rather than on a person's position or the position of another agent or object. For instance, members of the Guugu Yimithirr tribe will say, "The girl is standing north of the tree" instead of "The girl is standing on the left side of the tree." This implies (and is affirmed by Deutscher's study) that speakers of this language will have an excellent sense of direction and will also memorize and recall people, objects, and events based on the cardinal directions rather than their physical characteristics.

It is argued that native language not only affects our thinking in the presence or absence of a particular vocabulary, gestures, or a grammatical category, but also in the manner in which we perceive events in nonverbal event representations, such as pictorial sequences. In other words, language can affect our mental imagery and its signifier in semiotic media, like gestures and pictures. Vastenius et al. (2016) found a significant difference in the manner in which Swedish and Kurdish speakers order pictorial sequences. The Swedes preferred the Actor > Act > Patient or SVO (subject-verb-object) sequence more often than their Kurdish counterparts, who preferred the Actor > Patient > Act or SOV (subject-object-verb) order in the pictures. The predilection of the Swedish and Kurdish participants toward these pictorial sequences emerges from the grammatical construction of the subject, object, and verb in the Swedish and Kurdish languages.

A creative pedagogical expression of linguistic relativity was designed by Professor Uday Athavankar for his course *Product Semantics* (DE-727) at the Industrial Design Center, Indian Institute of Technology Powai, India. He would ask the class to close their eyes and imagine the image that comes to their mind for two different words—*svepakghar* and *kitchen. Svepakghar* is the closest equivalent of the English word *kitchen* in the native Marathi language. Invariably, students described the visual equivalent of these two words differently. *Svepakghar* was often described as a dimly lit space or a room filled with natural light in which a saree-clad woman is preparing food while sitting on the floor on a mud stove or *Chulhā* in an iron *kadhāī*. All cooking ingredients in small ceramic or steel jars are spread around her. There is no wash basin close to her. A mortar and pestle are lying close by. Behind her, there is an old wooden cupboard. The floor has dark tiles. On the other hand, students described *kitchen* as an artificially lit, bright, often white-colored room, where a woman is preparing food on a gas stove, with a smoke-hood on the top. A white basin is installed on one of the platforms. White cupboards can be seen on the walls and below the platforms. The woman is cooking in a nonstick vessel and is surrounded by several electronic gadgets like a mixer-grinder and a microwave. A refrigerator is in a corner. The dramatically different descriptions of the mental images triggered by words from native and non-native languages reflect the extent to which language shapes our perception and therefore also the design of our artefacts and spaces.

The worldview of an individual is shaped and represented in language. Humboldt asserts that the mind processes the world through language and not only shapes the worldview but also brings the world into being (as cited by Underhill 2009, 99). No wonder that Orientalists fought for the medium of instruction to be Urdu or Sanskrit in the Indian educational system instead of English, as Macaulay had proposed in 1835.[2] Orientalists feared the hegemony of Western thought on the Indian mind if the vernacular language was replaced by English. History, however, took a different course.

Like Deutscher, Boroditsky (2009, 2011) and Levinson & Haviland (1994) also claims that different languages may impart different cognitive skills; hence, speakers of different tongues may also think differently. Boroditsky has collected data on the effect of language on the speaker's perception from around the world: China, Greece, Chile, Indonesia, Russia, and aboriginal Australia. Her data show that native language plays an important role in shaping habitual thoughts such as our perception of colours, objects, space, and time; native language profoundly shapes the way we think and see the world. Boroditsky also claims that speakers of different languages differ in how they describe events and, as a result, how well they can remember who did what. English speakers tend to phrase things in terms of a person/people doing the thing. Speakers of Japanese or Spanish, in contrast, are less likely to mention the agent when describing an accidental event. In Spanish, one might say "Se rompió el florero," which translates to "the vase broke" or "the vase broke itself." Thus, the agent or the author of the action remains unknown in such a conversation, whereas in English the authorship is clearly established.

[2] These comments are noted in "Minute by the Hon'ble T. B. Macaulay," dated 2nd February 1835 and available at http://www.columbia.edu/itc/mealac/pritchett/00generallinks/macaulay/txt_minute_education_1835.html. "All parties seem to be agreed on one point, that the dialects commonly spoken among the natives of this part of India contain neither literary nor scientific information, and are moreover so poor and rude that, until they are enriched from some other quarter, it will not be easy to translate any valuable work into them. It seems to be admitted on all sides, that the intellectual improvement of those classes of the people who have the means of pursuing higher studies can at present be affected only by means of some language not vernacular amongst them. What then shall that language be? One-half of the committee maintain that it should be the English. The other half strongly recommend Arabic and Sanskrit. The whole question seems to me to be—which language is the best worth knowing? ...We must at present do our best to form a class who may be interpreters between us and the millions whom we govern—a class of persons Indian in blood and colour, but English in tastes, in opinions, in morals and in intellect. To that class we may leave it to refine the vernacular dialects of the country, to enrich those dialects with terms of science borrowed from the Western nomenclature, and to render them by degrees fit vehicles for conveying knowledge to the great mass of the population."

This insight into agency has an interesting correlation with the worldview of several Indian cultural designers. Vatsyayan (1996) believes that in Indian culture, the identity of the "self" or "I-nes" as unique and distinct from another is not important. The creators search not for the unique or the novel, but to situate themselves in an ongoing discourse that is larger than themselves (p. 2). In Varanasi, toymakers, irrespective of the source of origin or ideation of the toy design, claim, "*Haṁhīṁ to banāye haiṁ.*" (This translates as "We have made this.") If prodded further to know the original designer of the toy, the toy designers simply claim ignorance but persist in their claim on the design. In modern scenarios, this could result in a case of contested intellectual property (IP) rights or a loss of authenticity due to dubious authorship. However, the word *Haṁhīṁ* in Northern India embodies both a singular as well as a plural person at the same time. In English, this would imply that *Haṁhīṁ* denotes *I* and *we* simultaneously; therefore, instead of saying "I made this", *Bāṇārasī* toymaker literally says, "We made this" while essentially connoting a singular self. The absence of a name of the original authors or even their memory, besides the use of the first-person plural pronoun by all toymakers of the *Bāṇārasī* cultural design community, connotes the equitable ownership of the toy and distribution of its know-how among all. The toy designer is a grammatical subject that serves the function of the toy without alluding to the history, biography, or psychology of the original author. The historical, biographical, and psychological context is an inherited and shared identity not of an individual toy designer but of an entire community of parents and their offspring. In other words, the toy designers are born simultaneously in the family as well as in a professional community, where they inherit the capital, credit, and accumulated experience of both the father and their community (Thaplyal 1996, 55–56). Thus, community ownership is manifested in the individual agency of the toy designer in the *Bāṇārasī* language.

In summary, according to the theory of linguistic relativity, in a community, language coordinates the perceptions and actions of its speakers while enacting their conception of the artefacts they use. The meanings that artefacts acquire in use are often framed in the language of a particular community well before they are actually used by people as social or cultural artifacts. Therefore, meanings of artefacts in language denote how artefacts come to live in the narratives of their speakers, in effect making the artefacts available or unavailable for use. Thus, the fate of all artefacts is decided in language (Krippendorf 2006, 148–49). If so, can we capture the cultural meaning of the *Bāṇārasī* toy from the *Bāṇārasī* language?

The word *toy* does not imply the same object in all languages, nor does its meaning remain constant over time. This cultural notion of toys is beautifully articulated in the paper of Kyburz (1994, 1–28), who differentiates the Japanese word *Omōcha* from its English counterpart, *toy*. The purpose of toys in the Western sense lies in their practical function is ludic, whereas that of *Omōcha* lies in its meaning. As practical function gives way to symbolic meaning, *Omōcha* turns into a symbol, moving out of the child's hands and into the conceptual realm of adults. It is concerning this class of toys, which lack all ludic purpose in form as well as intent, that Kyburz attempts to position the definition of toys in a cultural context. The essential formal characteristic of the *Omōchas* is their miniaturized representation, in figurative terms, of certain human beings, animals, or material objects expressive of symbolic values that do not intrinsically relate to play, games, sport, amusement, or entertainment. They are refined objects of aesthetic appreciation for adults, items of the collection, or simply souvenirs brought back from a journey. But the Japanese traditional concept of *Omōcha* appears to have died in the latter half of the Meiji Restoration with the advent of the industrially produced "Western toy."

Examples of adult toys, such as *Omōchas*, have been a part of several cultures at different times, sometimes even as a part of a royal collection. Tipu Sultan, the ruler of Seringapatnam, India, kept a wooden automaton (famously called Tipu's toy) in his personal room. This toy is a wooden automaton that imitates the growling sounds of a tiger devouring a British soldier. The sounds of the tiger killing the helpless British soldier had a cathartic effect on the Sultan's desire to vanquish the British. According to the seventh descendant of the Sultan, Syed Mansoor Ali, the Sultan used the toy to calm himself in between the continuous armed conflicts with the British, and he found vicarious pleasure in the death of the wooden British soldier.

2 From *Kṛidānakā, Khilōnā* to a Toy: An Etymological Overview of Artefacts

The Dutch historian Huizinga has written extensively on the notion of play, so much so that he titled his book *Homo Ludens*—man, the player (1950). Huizinga postulates that every language, in forming its idea and expression of play, has found a different word to describe it. Words define and perhaps limit the use of an idea. In the plethora of understandings of what play could possibly mean, Huizinga summarized it as an intrinsic activity that exists for its own sake, outside of ordinary life. This status of play and its consciousness outside the ordinary life of the player is similar to the Indian concept of लीला (*Līlā*). It is God's spontaneous and joyous expression of a purposeless world (Aleaz 2004, 4–5). *Līlā* is divine play: Gods play the world into being outside of themselves. Similarly, *Raslīlā* (where *ras* means "aesthetics" and *līlā* means "act/play"; therefore, *Raslīlā* means "play of aesthetics") is also considered a divine sport played by Lord Kṛṣṇa with his consort Radha and other *Gopīs* (milkmaids) on the occasion of *Janmāṣṭamī* (a Hindu festival that celebrates the birth of Lord Kṛṣṇa) on the banks of the river Yamunā in Vrindāvan. Therefore, it is often said by the *Bāṇārasī* community and also Hindus in general that Gods are *playing* their *Līlā*. Huizinga, who was also proficient in Sanskrit, speaks of two more verbal roots for the play concept in Sanskrit. *Kṛidāti* denotes the play of animals, children, and adults. *Divāyti* connotes gambling, dicing, and joking, jesting, or mocking another.

Kṛidāti and *Kṛidānakā* (literally a "plaything" or a "toy") share the same root that is *Kṛīd*. *Kṛīd* denotes play and is also a root for many other kinds of activities ranging from dalliance to a pleasure hill (Macdonell 1979). Monier Williams (1899, 321) traces several meanings of both the words—*Kṛīd* and *Kṛīdā*—as discussed in the following definitions.

Kṛīd

Kṛīdishyat, to play, sport, amuse oneself, frolic, gamble, dally; to jest, joke with : Caus. *Kṛīdāyati*, to cause to play, allow to play; *Kṛīdā*, m.f.n., playing, sporting (said of the winds); (as), m. sport, play; f. sport (often in competition., e.g. *krisa-mudah*, f.pl. the pleasures of playing amorous sport); Kṛṣṇa, sport with Kṛṣṇa; *jala-k*, playing about in water.

Kṛīda

Kapit-tva, n. imitating an ape for amusement or in jest; -*Kān*ana, n. a pleasure grove; -*Kasāra*, a pleasure-pond; -*Kumāra*, m., N. of a *Kumāra* or a *Gandharva, Balar*; -*K<u>u</u>ta* (*Kṛīdāka*), n. sportive or wanton purpose or desire; -*Ketana*, n. a pleasure house, *Kathas*;

-*Kopa*, m. anger in sport, assumed anger; -*Kautuka*, n. wanton curiosity, sport, pastime, enjoyment, lasciviousness; -*Kausala*, n. the art of joking; -*Khanda*, n.N. of GanP.; -*griha*, m.n. a pleasure house; *Cankramana*, N. of a locality; -*Candra*, a meter of 4 × 18 syllables; -*Tāla*, m. (in music) a kind of measure; -*Narī*, f. a harlot, courtesan; *Paricchada*, m. plaything, toy; -Parvata, m. a playhill or pleasure hill, pleasure mound, or hillock in a garden; -*Parvataka*, m. id., ib.; -*pura*, n. a town visited for amusement; -*Pradeśa*, m. a playground;

-*Maya*, mf(i)n. consisting of play or sport; -*Mayūra*, m. a peacock kept for amusement; -*Markata-puṭa*, m. a young monkey serving as a plaything; -*Mṛiga*, m. an animal kept for pleasure, toy deer;

-*Ratna*, n. gem of sports, copulation; -*Ratha*, m. a pleasure chariot, carriage used for amusement (as opposed to a war chariot); -*Rasa*, m. enjoying sport; -*Rājata-śṛṅgārapatra*, Nom. To represent a silver liquor-cup used as an object of amusement;

-*Rudrāya*, Nom. To resemble *Rudra* while playing; -*vat*, mfn. sportive, playful; -*vana*, n. a pleasure grove, park; *Vāpi*, f. a pleasure pond (in which turtles and fish are kept for pleasure); -*vesman*, n. a pleasure house; -*sakunta*, m. a bird kept for amusement; -*saras*, n. a pleasure pond; -*Sthāna*, n, a playground.

Sastri (2003) divided the Sanskrit word *Kṛīdā* in his *Sanskrit to English Dictionary* into two sets. He speaks about *Kṛīdā* that is performed by *Śāmnyājan* or the common folk, which is similar to the idea of play with a ball, sport, or pastime jokes. Along with this, he also mentions 22 other kinds of *Kṛīdā* that are practiced by the *Vilaśjñān* or elite folk, as mentioned in the ancient Sanskrit text *Sarasvatī-kaṇṭhābharaṇa*. These 22 kinds of adult play were largely played by women according to the seasons of the Hindu calendar. These play types were associated with religion, festivals, and the seasons, and included a wide variety of activities ranging from fasting to playfully hitting each other with flowers, playing *Holi*, and bathing and boating during the summers in the river. The adults were important participants in this meaning of *Kṛīdā*, unlike the play for *Śāmnyājan*, which included children. The adult play required very limited artifacts and instead had a strong component of pretend play (refer to Table 1).

Table 1 *Krīḍā* practiced by *Vilaśjñān* mentioned in *Sarasvatī-kaṇṭhābharaṇa*

No.	Type of play	Play
1	Games	*Āñdolan Caturthī*: Playing on swings during the rainy season. Under the *Śāmal* tree, women play blindfolded games
2	Ritual	*Āṣṭamī Cāndraka*: In the month of *Caitra*, women break their fast after making an offering to the moon. *Madanotsava*: On the eve of *Holi*, prayers are offered to *Kāmadeva*
3	Feasting	Sweets are kept under the rays of the full moon of *Śāradā Pūrñimā* for the whole night and eaten in the morning. *Abhyuṣkhaḍika*: Raw or roasted green peas from the farm are eaten. *Ikṣukhaḍika*: Freshly cut sugarcane is eaten
4	Dalliance	*Chutbhājikā*: Plucked flowers of the mango tree are offered to *Kāmadeva* and then used to decorate the hair *Chutlatikā*: With the stems of a plant, women hit each other and tell their lover's name. *Biskhadikā*: Lovers pluck the lotus from the ponds and eat it together *Kautuka*: Wanton curiosity *Gṛhā*: Pleasure house *Kanana*: Pleasure grove
5	Leisure	*Kanda Caturthī*: Women spread barley on the floor and sleep on it *Puṣpakyācika*: Women pluck the flowers of the *Maulsarī* tree and place them in a wine cup *Kasāra*: Pleasure pond *Parvata*: Artificial pleasure hill
6	Pretend play	*Aśokottasika*: Rich women dress up and theatrically hit the *Aśoka* tree in a bid to make it grow higher *Navarātrikā*: After the first rain, girls play pretend marriages on the lawn *Kopa*: Feigned anger for fun
7	Adornment	*Suvasantaka*: Welcoming the spring with yellow clothes and flowers
8	Sports	*Toyakrīḍā*: Swim in the river during summer
9	Celebration	*Yakṣarātrikā*: To celebrate *Dīpāvalī*—a Hindu festival *Udakṣevādika*: To throw water on each other with a water gun during the festival of *Holi*
10	Spectatorship	*Prekṣā*: To watch a play with the family
11	Jest	*Kadambyūḍha*: In the rainy season, using the stalks of tender sugarcane or flowers, friends hit each other playfully *Bhūtmātṛkā*: One friend cajoles the other friends in a group *Kapitavya*: Jesting imitation of a monkey *Kauśala*: Art of jesting

2.1 **Khilōnā** *to a Toy*

Despite the rich semantic signification of *play* and *toy* in the Sanskrit language, there is no sign of it in the actual practice of toy making or even playing by adults or children in India today. We cannot help but speculate that the death of the Sanskrit language itself, in addition to the avalanche of changes that the Indian society went through after independence through industrialization and liberalization, may have something to do with this phenomenon. Sanskrit is not used in common parlance today; nor does a category of adult play, along with its 22 manifestations, exist in India in general, or in contemporary Bāṇāras. Sanskrit was replaced by Hindi and Urdu in day-to-day communication in the colonial days, and after independence, Hindi was declared the national language of India.

In Hindi, toys are called *Khilōnā*, a derivative of the word *Khel*, which means "to play." *Khel* includes some meanings that are close to the Sanskrit words mentioned before; for instance, *Līlā* or the categories of adult play mentioned in the *Sarasvatī Kaṅkābharaṇa*, such as rituals and ceremonies. The Sanskrit words are no longer used, but their meaning is still visible in the activities considered as play in Hindi. Bahrī (2009) refers to these meanings of *Khel* and *Khilōnā* in his dictionary.

Khel—masculine

1. Activities are done for relaxation and pleasure.
2. *Kartāb*—feat or a trick, for instance, acrobats in a fair.
3. *Tamāśā*—a show, an exhibition, stunts, for instance, the monkey man, a dance performance, or acrobats.
4. *Abhinaya*—acting, performance, for instance, a *Nāṭaṅkī* (popular folk theatre performance), or *Vidūṣaka* (jester).
5. *Līlā*—spectacle, game show, for instance, the killing of Rāvaṇa in *Rāmlīlā* (a dramatic folk performance of the life of Lord Rāma) or *Nāgnāthīya* (a theatrical enactment of the story of Kṛṣṇa killing the serpent Kaliya Mardana).

Khilōnā—feminine

1. Objects of play such as a wooden toy.
2. Objects of entertainment.
3. Very cheap objects.

Although the Hindi meaning of *play* and *toy* is narrower than the Sanskrit term *Kṛd*, it nevertheless establishes the child as an important user of the *Khilōnā* while also referring to its ludic character. The artefact takes a significant role in these definitions as it is characterized by various attributes such as "wooden toy, cheap and entertaining." The adult in this meaning becomes either a participant in creating an object or a spectator of performances such as the *Nāuṭaṅkīs*. But, unlike the *Sarasvatī Kānthābharan*, no specific adult play is mentioned here. According to this set of meanings, *Khel* refers not only to sports such as *Gilli-Daṇḍā* (a game with a stick and a ball) or *Kushtī* (wrestling) but also to performances and representation of stories. Huizinga (1950) describes representation as a function of play. *Tamāśā*, *Abhinaya*, *Līlā*, and *Kartāb* are all representations of performances; very often of a story. For instance, *Nāuṭaṅkīs* in Northern India enacted during *Rāmlīlā* are amateur folk theatres with elaborate sets narrating the life of Lord *Rāma*. *Nāga Nāṭhāīyā* in *Bāṇāras* is the theatrical representation of the child *Kṛṣṇa* dancing on the head of a venomous serpent, *Kāliya*, in the river *Gaṅgā* itself. This annual performance is held at the *Tulsi Ghat* on a wooden head of the serpent floating in the river, upon which a young child dressed as Kṛṣṇa is seen dancing. According to Hindu mythology, a young Kṛṣṇa defeated the mighty *Kāliya* in a fierce underwater fight and then danced on the head of the vanquished. The narrative is portrayed not only in the larger-than-life performance in the river but also in wooden tableaus (refer to Fig. 1).

These religious performances are also called *Līlā,* which involves *Abhinay* of actors; but Aleaz (2004) differentiates it from its Sanskrit meaning discussed earlier. These religious performances or dramatizations are theatrically staged, played by actors unlike the divine play of the "supreme" referred to in Sanskrit as *Līlā. Līlā* played by the "supreme" is enacted by human actors in what they would consider as reality.

This understanding of *Khel* as a folk and religious extension of performance and display broadens the number and kind of artifacts that can be accepted in extension as a *Khilōnā*. Some of the objects, such as elephants, kings, and queens, allow children to engage in pretend marriages, processions, and other celebrations (Roopnarine et al. 1994, 19–24). These toys, used for both pleasure and education in addition to performance rituals, are not similar to the wooden toys that perform a ludic function alone. The meaning of *Khilōnā* here embodies an enculturative mechanism. That is, through *Khelna* or play children learn societal roles, norms, and values (Schwartzman 1978). *Bāṇārasī* toys are a good instance of enculturative *Khel* and *Khilōnā*. Here, children may rely equally on stories as a medium of play as well as a stimulant for their imagination (Roopnarine et al. 1994, 23). In Bāṇāras, stories and storytelling are an integral part of life. Myths and stories centered around divine figures, as written in the *Purāṇas* and epics like the *Rāmāyaṇa* and *Mahābhārata*, form the main source of representations in *Khilōnās*, especially the carved statues.

Khilōnās are representations of stories.

Transcription 2.1: Jeeravati Devi, a local resident of Bāṇāras

These stories are read in various holy texts or heard during community recitations of religious texts such as *Paaths*, *Pravachans*, or *Katha-kathan* by *Pandits* or scholars. They are also told through performances staged during the numerous *Melas* (fairs) of Bāṇāras. The representation of carved statues of deities draws directly from these texts, with gods and mythological figures depicted according to detailed descriptions.

For instance, Viṣṇu and Kṛṣṇa are portrayed with blue bodies, and the eyes of living beings are often shown as *Kamalnayānī* (lotus-like). Brahmā is depicted with three heads, while Rāvaṇa is represented with ten. Gods are commonly depicted as couples rather than individuals, such as *Rādhā-Kṛṣṇa*, *Śiva-Pārvatī*, or *Viṣṇu-Lakṣmī*. This practice also extends to the mythical vehicles, or *Vāhanas*, assigned to each deity—Lord Gaṇeśa with a rat, Lord Śiva with Nandi the bull, and Lord Kārtikeya with a peacock.

These visual and symbolic cues are deeply rooted in the stories narrated in the holy texts and are a vital part of the *Bāṇārasī* culture, whether shared at home or in a *Pāṭh.* The practice of storytelling through religious narratives and symbolic representations strengthens the connection between traditional toy designs and their cultural significance. As Krippendorf (2006, 169–71) states,

(a)

(b)

Fig. 1 **a** Celebration of *Nāga Nathaiyā* in Bāṇāras. **b** Representation of *Nāga Nathaiyā* in a wooden tableau by *Bāṇārasī* toy designers. *Image source* **a** Retrieved https://commons.wikimedia.org/wiki/File:Nag_Nathaiya_festival_in_Varanasi.jpg under the Wikipedia (open source) license of the photograph taken by Nandan Upādhyāya. This file is licensed under the Creative Commons Attribution-Share Alike 3.0 Unported license. **b** Author

> Humans live in story. Humans explain through story. Humans explore their world through story. Humans design their worlds as stories. Individuals are born into the stories until they die. Similarly what one knows of an artifact is the stories one has heard and one tell of it. The artifacts that people surround themselves with mean the very roles they play in the stories in which they occur.

Parents in *Bāṇāras* believe that *Khilōnās*, which are seeped in customs and religion, teach children not only beautiful stories but also infuse them with values. *Bāṇārasī Khilōnās* enact stories on tableaus or *Jhāṅkī* that are made and displayed in every house during festivals. *Jhāṅkī* is a *Līlā* playfully created by mortal humans instead of the Gods. It is a temporarily real world of its own in which a sacred performance is played, performed, or created (Aleaz 2004). Most statues of deities, especially those of Lord Kṛṣṇa, have become an element of decoration used to enact a story; these statues are handled by children and adults alike during the festival of *Janmāṣṭamī*, which is the celebration of the birth of Lord Kṛṣṇa. Stories of Kṛṣṇa dancing with the *Gopīs*, *Pūtnā vadh*, and *Vāsudeva* with the baby Kṛṣṇa sheltered by a serpent are mostly enacted through wooden *Khilōnās* in Bāṇāras. (Refer to Transcription 2.1 and Fig. 2.) Children help their parents during the festivals by displaying episodes of mythical stories in elaborately designed *Jhāṅkīs*. Here, *Khelnā* is an activity not just between child–object and child–child, but also between a child and an adult, where the child and the adult can be both a participant and a spectator.

Seema: Look at this. This is the *Dāṇḍīyā Rās* (a form of dance in which Lord Kṛṣṇa dances with his *Gopīs*). This is a big boat. In this, Lord Kṛṣṇa and *Rādhā* will go for a ride in the *Gaṅgā*.

Author: Where is the Sakha in this?

Seema: The Sakha is not here now. The Sakhi is playing *Dāṇḍīyā* with God. And this tree here is also made of wood. This here is a peacock. (Sakha and Sakhi are friends of Kṛṣṇa.)

Author: What is this wooden tree called?

Seema: This is called a tree only. It is the Kadamba tree on which God used to play. And, here is a boat, a cow. This is God. It is such a beautiful statue. This is a jail here—a small jail where God was born.

Transcription 2.2: Seema Agarwal, Housewife residing in Vishwanath Gali, Bāṇāras

Note the use of the word *Khelna* or *play* in two different contexts in Transcription 2.2. According to Seema Agarwal, *Dāṇḍīyā*—a dance form—is not performed but is played by the Sakhi with God. *Dāṇḍīyā* is a part of the *rāsa-līlā* in which Lord Kṛṣṇa is said to have danced with the *Gopīs* or milkmaids. Every *Gopī* believed that Kṛṣṇa was dancing with her, but that is how God played his *līlā*. In the second instance, Lord Kṛṣṇa, as a small child, played on the Kadamba tree. This play of a child is ludic play and is not associated with divinity.

Festivals such as *Janmāṣṭamī* or *Rām-līlā* associated with Kṛṣṇa and Rama are periods of intense activity for the toy designers. The local market in the Vishwanath Gali is filled with local residents desirous of purchasing wooden *Khilōnā*s for their

(a) (b)

(c) (d) (e) (f)

(a) Kṛṣṇa lifting Mount Govardhan (b) Kṛṣṇa and *Rādhā* in a boat ride of the river Gaṅgā (c) *Vāsudeva* carrying baby Kṛṣṇa protected by the Nag (d) Milkmaid (e) Lord Viṣṇu (f) Lord Gaṇeśa

Fig. 2 Sample of narrative toys displayed in a *Jhāṅkī*. *Image source* **a–f** Author

children or for a ritual; these festivals are a profitable time for the toy designers. *Janmāṣṭamī* is a Hindu festival that celebrates the birth of Lord Kṛṣṇa. These wooden statues of deities are considered objects of play and not sacrilege because the rituals of their *Sthāpanā*, or installation in the temple, have not been performed. Nonetheless, these statues are handled with respect because they do not serve a ludic purpose alone. (Refer to Transcription 2.3.) If the statues break or are disfigured for any reason, it is believed that they have become *Khaṇḍit* or defiled and therefore are inauspicious. The *Khaṇḍit* statues are offered to the river Gaṅgā. This is not the case with objects of decoration or ludic play, which are simply thrown away if they break. Dongerkey (1954) says that a child playing with toys representing religious personalities or characters learns to handle them with great care because of the reverence and respect they evoke. The child's destructive tendency can be checked by religious subject matter.

These *Khilōnās* are meant for decoration, but they are also given respect. We are careful not to touch them with our feet, or that they are not simply lying around on the floor. Plastic toys can be easily given to the children.

Transcription 2.3: Jeeravati Devi, Housewife residing in Gadolia, Bāṇāras

Most *Khilōnā*s used for *Janmāṣṭamī* are removed on *Chattī*. *Chattī* is the sixth day after a child is born, during which the child is given a name. After *Chattī*, *Khilōnā*s are neatly packed and stored for the next year, to be opened and played with during *Janmāṣṭamī*.

These ritualistic *Khilōnā*s made for festivals differ from each other according to the story and the festival celebrated in Bāṇāras. During *Rām-līlā*, various masks of gods and goddesses are worn by children, who play in the streets enacting the story of the triumph of good over evil with bows, arrows, and swords made of bamboo or wood. This *Abhinaya*, since it is played on the streets, is outside ordinary life and is enacted within a physical and temporal boundary according to the rules of the mythical story. The story is played by actors rather than statues, with the help of *Khilōnā*s like the bow and arrow.

As festivals change, so does the production of *Khilōnā*s; thus various forms of play are seasonal in Bāṇāras. The local market does not feature clay *Khilōnā*s of animal figures after July because *Janmāṣṭamī* is celebrated in the month of *Māgha*, which is August. During *Diwālī*, clay *Khilōnā*s are generally sold. For *Nāgapañcamī*, clay *Khilōnā*s of the snake charmer, along with the *sugga* (parrot) and the pigeon, are seen in the market. Aleaz (2004) relates religious festivals with play because both transcend ordinary life; mirth and joy sometimes dominate in both. The festivals are limited by time and place and combine strict rules with genuine freedom.

Although the category of adult play mentioned in the *Sarasvatī-kaṇṭhābharaṇa* no longer exists, it is interesting to observe vestiges of ritual (*Aṣṭamī-candraka, Mādanotsava*), dalliance (*Chutbhājikā, Chutlatikā, Biskhadikā, Kautuka, Gṛhā, Kanana*), and celebration (*Yakṣa-rātrikā, Udakṣevādikā*) in symbolic forms in *Bāṇārasī* ceremonies for adults. Although no Hindi equivalent of these Sanskrit terms is found in contemporary vocabulary, symbolic gestures representing these forms of play are nevertheless enacted in the ceremonies.

For instance, in the celebration of marriage, the newlywed bride is presented with *Khilōnā*s in the ritual of *Bhārai*. *Bhārai* is an auspicious offering or gift. A set of *Band-bājā* (music players) along with other *Khilōnā*s like *Chusnī* (pacifiers) and *Jhunjhunā* (a rattle), are included in this ritual as symbols of fertility. The wooden *Sindorā* (a small box used to store vermilion or *sindoor*, which married women apply on their foreheads as a marital symbol) is also given as a gift, wrapped in a red muslin cloth. A *Gudā* (a male doll) is specially offered as a sign of the birth of the first son. Recently, *Samdhi Kā Khilōnā*, an erotic toy, has also become part of this ritual for fun and laughter. Interestingly, this covert play, used to tease and embarrass the bride, is similar to the adult category of dalliance mentioned earlier. However, such instances are rare and not expressed with a specific word in the vocabulary.

Seema: These toys are sent by the in-laws to the bride's family, [and we call these toys] *Hathpudi*. In the *Hathpudi*, apart from sarees and jewellery, there are also toys.

Father-in-law of Seema: And the doll also goes at this time. But this doll is a male doll. After the marriage, during the night, this ritual is carried out.

Seema: This ritual is called *Bhārai*. In this ritual, all toys are put in the lap of the bride and then she is sent off to her new home.

Father-in-law of Seema: In order to have the first child as a son, the male doll is given. Everything is sent—sweets.

Author: So what are the kinds of toys sent to the girl apart from the male doll?

Seema: They are all related to children. There is one special toy called the *Samdhi kā Khilōnā*. There is a spring inside the cylinder from which something pops out. That's a fun kind of toy. Then there are toys like the rattle.

Transcription 2.4: Seema Agarwal—Housewife residing in Vishwanath Gali, Bāṇāras

Apart from the Vilasijan, the *Sarasvatī-kaṇṭhābharaṇa* also referred to play for the *Saṃnyājan*, where children were dominant users alongside the common adults. The Sanskrit word *Kṛdānaka* denotes objects intended for children. Unlike the play of the Vilsijan, play for the *Saṃnyājan* was primarily ludic and often lacked religious or moral undertones.

While ancient Indian literature doesn't extensively cover children's play, evidence of such play exists in various sources. For instance, remnants of toys like bullock carts have been discovered at Harappa and Mohenjodāro sites. Temple panels from Amravati, dating back to the second and third centuries CE, depict children with toys such as toy horses, toy elephants, two-wheeled carts, and rattles. These artifacts suggest a rich tradition of play with both symbolic and functional aspects (Dongerkey 1954, 38–42). Additionally, miniature paintings reveal a range of toys used for play, including rattles, push and pull toys, and spinning tops.

Ancient literature also mentions various kinds of toys, including automata. For example, the *Kathāsaritsāgara* ("Ocean of the Streams of Stories"), written by Somadeva in the eleventh century, includes tales of mechanical dolls and magical contrivances. In Book VI, Chapter 29, Somaprabhā uses mechanical wooden dolls and magic machines to entertain her friend Kaliṅgasenā. Chapter XLIII describes Rājyadhara, a carpenter who designed wooden automata shaped like humans to populate an empty city and made it appear as if the automata were alive.

This ludic nature of the toy became popular with the wide acceptance of the English word *toy* through British colonization. Consequent to the Macaulayan policy, the English language, though unfamiliar to Indian culture, was used as a medium of instruction in schools, and even today is widely written and spoken. Many English words have assimilated into vernacular languages; for instance, *decoration, fun-type*, and *specially* are used without any change in Hindi by *Bāṇārasī* residents.

Interestingly, similar to the Sanskrit word *Kṛd*, the English word *toy* also connoted more meanings of play than a ludic one alone in its past usage. *Toy* comes from the Middle English (1100–1450) *toye* (Toy, n.d.). It referred to *amorous play* until about the 1700s; the meaning can be broadened to include dalliance and fondling. It is interesting to note that the word *toy* was largely associated with adult pleasure for at least six to seven centuries. By the 1500s, the meaning of the word *toy* shifted into the realm of the child—*piece of fun or entertainment* (1500), *thing of little value, trifle* (1520s), and *thing for a child to play with* (1580s). By the seventeenth century, the *toy* attained the ludic meaning, which is currently in use—*something to play with, especially as intended for use by a child* (Longman 1984, 1590; Onions et al. 1966, 934).

Of course, adult users and adult use of toys are springing back to life in contemporary times. Fidget spinners were designed not only for marginalized users such as children with ADHD, but also for anxiety-stricken adults. Toys have changed their function (e.g., STEM toys, toys for healing post-traumatic stress disorder), meaning (stress busters, cultural identity), users (astronauts, convalescing patients), value (expensive collectors' items), as well as space of use (war-torn zones) in contemporary times.

However, the toy industry continues to center around the child as its primary user. Referring to the Oxford dictionary, Kyburz (1994) discusses the Western sense of a toy as a material object for children or others to play with, contrived for amusement rather than practical utility. This meaning of *toy* has found wide acceptance among both Western and Eastern urban consumers; the global toy and games market is bereft of cultural notions. Likewise, in *Bāṇāras*, too, the English meaning of the word *toy* in the ludic sense for children has been accepted, but some of the older meanings of *Khilōnās* have not been discontinued. Today, children help their parents during the festivals in making the tableaux of mythical stories but have also gradually shifted from ritualistic traditional *Khilōnās* to modern plastic and soft toys in everyday life, thereby relegating the cultural meanings of *Khelna* to specific and special times, places, and occasions alone. The word *Khilōnā* itself has also acquired ludic adjectives indicating objects of play and entertainment artefacts that are often very cheap, as mentioned in the definition of *Bahari* above. This ludic connotation of play is also visible in some of the *Bāṇārasī Khilōnās*. These *Khilōnās* are not similar to the ritual statues used in the *Bāṇārasī* festivals and ceremonies. *Khilōnās* made for ludic purposes alone are often miniaturized and simplified representations of reality rendered in bright colors. Examples include the helicopter, *lattu* (spinning top), Cugṭī Ciṛiyā (pecking toy), *Hilānta* (a single pecking animal), Russian nested dolls, *sukhī parivār* (a *Bāṇārasī* nested doll), a set of musicians, and push and pull animals and birds, among others (refer to Fig. 3).

Edutainment toys that aid children in coping with their school curriculum are conspicuous by their infrequency or absence in *Bāṇārasī* toys. This may be because *Bāṇārasī Khilōnās* largely serves an enculturative role rather than an educational one, as discussed previously. A few toys, such as *chusnī* (pacifier), *jhunjhunā* (rattles),

(a) Rocking horse (b) Pull toy (c) *Hilānta* (d) Spinning top

Fig. 3 Samples of *Khelne Kā Khilōnā*. *Image source* **a–d** Author

nesting toys (joker toy), and walkers associated with the gross motor skills of children have, over time, been replaced by cheap plastic equivalents.

In summary, over time, many attributes of the meaning of *Kṛīdā*, *Kṛidānakā*, *Khel*, and *Khilōnā* in both Sanskrit and Hindi have ceased to be a part of daily conversations, such as the Sanskrit understanding of adult play. On the other hand, some parts of the older lexicon and its meanings have also been retained. For instance, the representative nature of play through performances, *Līlā*, and rituals still forms an integral part of the *Bāṇārasī* lexicon and practices. The ludic meaning of *Khilōnā* has been held in common by all three languages—Sanskrit, Hindi, and English. Despite the widespread use of English and the Westernization of the Indian lifestyle, the English word *toy* has not completely replaced the vernacular vocabulary and its meanings; however, its growing influence also cannot be ignored. This influence is evident in the proliferating use of Barbie dolls, Legos, and computer games in the burgeoning toy market, which is nevertheless strictly demarcated by the local *haāts*, *melās*, and shops that sell the wooden *Khilōnās*. The change in *Bāṇārasī* user preferences and the worldview toward ludic play is a result of not just market dynamics, but also of heavy advertising, the hegemony of modern designs, and the natural decline of several forms of cultural play.

As discussed previously, some meanings have been retained over time in the *Bāṇārasī* lexicon while others have been assimilated or become redundant. As meanings died in *Bāṇārasī* language, so did their physical manifestations, as evident in the absence of the performative, amorous, and ritualistic roles of adults in the *Kṛīdā* of the *Vilaśjñān.* The semantic change of the notion of play, as evident in the narrowing of its meaning from Sanskrit to Hindi and then to English, is marked by overlaps, alterations, replacements, and elimination of notions of play and toys, as represented in the Venn diagram in Fig. 4. Circle 1 encloses the Sanskrit words for play and play objects. Circle 2 encloses Hindi words, and Circle 3 encloses words in English. In Circle 1, all words denoting adult play are no longer in use and, therefore, do not overlap with either Circle 2 or Circle 3. On the other hand, the *Śāmnyājan Kṛīdā, Līlā*, and *Kapit-tva* are shared with the Hindi lexicon in Circle 2. *Nautankī* and acrobatics are exclusive to Circle 2, but they share with Circle 3 a growing emphasis on the child as the primary user of play. Circle 3, on the other hand, also has many words that are limited to it alone, such as *edutainment* or educative toys. The only meaning shared by all three languages is that of ludic play, as evident in the overlapping of the three circles in the center.

The market drive toward the modern Western meaning of *toy*, autonomous of the cultural situation of its consumption, is yet to completely replace the *Bāṇārasī* notion of *Khel* and *Khilōnā,* which is specific to its worldview and expressed verbally in its vocabulary and notions apart from its distinct physical manifestations in the wooden artifacts. Most *Bāṇārasī Khilōnās* even today are related to the cultural experience of children, and the toy designers exclude them from the Western worldview of the ludic notion of play alone. This culture-specific meaning of toys as extended to rituals and ceremonies, seasonality, and local artisanship of locally sourced materials broadens the definition of *Bāṇārasī Khilōnā* from ludic experience alone to a cultural representation. The cultural meaning of *toy* is contextually specific to *Bāṇāras* as manifested in its language and its artefacts. Therefore, henceforth in this text (unless otherwise specified), *Bāṇārasī* toys will be referred to as *Bāṇārasī Khilōnā* to distinguish them from their English counterparts.

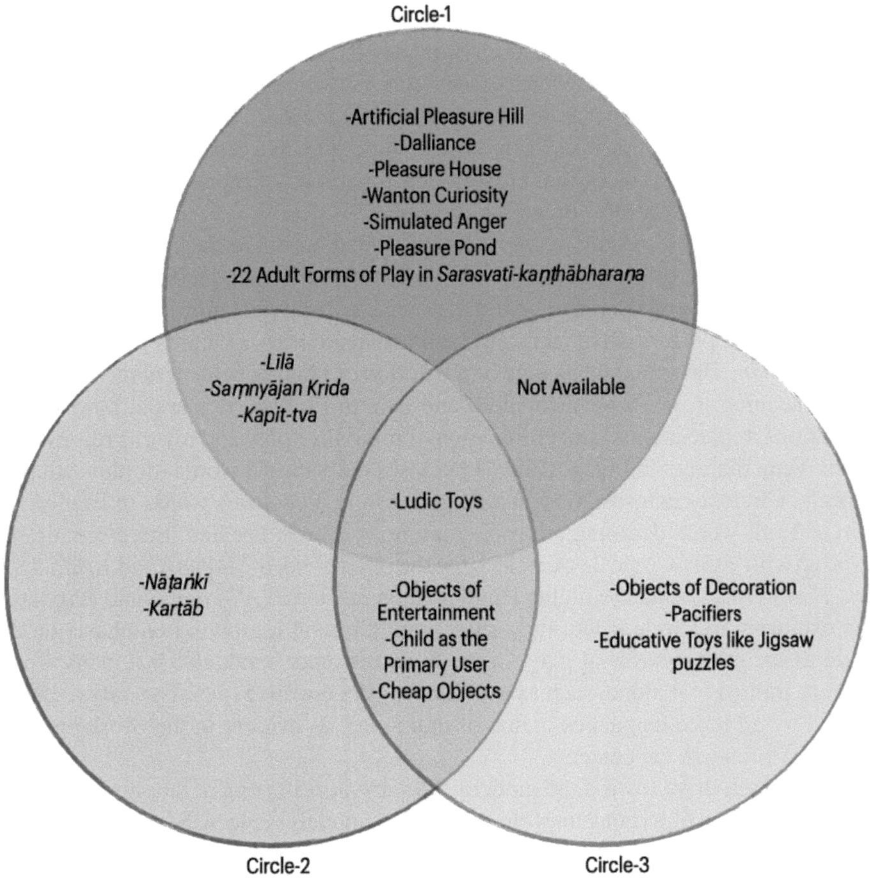

Fig. 4 Venn diagram of changing *Bāṇārasī* lexicon of play and play objects. *Image source* Author; it was also reproduced earlier in Patil (2024)

References

Aleaz KP (2004) Play and religion: indication of an interconnection. J Asian Res Center Religion Soc Commun 2(1):79–90

Bahri H (2009) Hindi Shabdkosh. Neha Publishers & Distributors, New Delhi

Boroditsky L (2009) How does our language shape the way we think? In: Brockman M (ed) What's next: dispatches on the future of science. Vintage Books, New York, pp 116–129

Broditsky L (2011) How language shapes thought: the languages we speak affect our perceptions of the world. Sci Am 63–65

Deutscher G (2010) Does your language shape how you think. New York Times 29

Dongerkery KSK (1954) A journey through toyland. Popular Book Depot, Bombay

Hoijer H (1954) The Sapir-Whorf hypothesis. Lang Cult 4:92–105

Huizinga J (1950) Homo Ludens. The Beacon Press, Boston

Krippendorff K (2006) The semantic turn: a new foundation for design. CRC Press, London

Kyburz JA (1994) "Omocha": things to play (or not to play) with. Asian Folklore Stud 53(1):1–28

Levinson SC, Haviland JB (1994) Introduction: spatial conceptualization in Mayan languages 613–622
Longman (Firm) (1984) Longman Dictionary of the English Language. Longman House, London
Macdonell A (1979) Arthur, Sanskrit English Dictionary. Award Publishing House, New Delhi
Monier-Williams M (1899) A Sanskrit-English Dictionary. Clarendon Press, Oxford
Onions CT, Friedrichsen GWS, Burchfield RW (eds) (1966) The Oxford dictionary of English etymology, vol 178. Clarendon Press, Oxford
Patil K (2024) Shape of the design worldview: does language inform the design sense? Design Issues 40(3):88–104
Pound E (1934) Make it new. Faber and Faber, UK
Roopnarine J, Hossain Z, P. Gill, and H. Brophy (1994) Play in east Indian context. In: Roopnarine J, Johnson J, Hooper F (eds) Children's play in diverse cultures. State University of New York, Albany, pp 9–31
Sastri S (2003) Ashok Sanskrit-Hindi-English Dictionary, 1st edn. Ashok Prakashan, New Delhi
Schwartzman H (1978) Transformations: the anthropology of children's play. Plenum Press, New York
Thaplyal KK (1996) Guilds in ancient India. New Age International Ltd, New Delhi, p 180
Toy. Online Etymology Dictionary (no date). Retrieved from: Dictionary.com: http://dictionary.reference.com/browse/toy. Accessed 30 Aug 2013
Underhill JW (2009) Humboldt, worldview and language. Edinburgh University Press, Edinburgh
Vastenius A, van de Weijer J, Zlatev J (2016) The influence of native language word order and cognitive biases in pictorial event representations. Cognit Semiot 9(1):45–77
Vatsayan K (1996) Bharata, the Nāṭyaśāstra. Sahitya Akademi, New Delhi
Whorf BL (1944) The relation of habitual thought and behaviour to language. ETC: Rev Gen Semant 197–215

Chapter 3
Sorting the *Bāṇārasī* Mind: What Do Linguistic Categories Reveal About *Bāṇārasī Khilōnās*?

1 Questioning the *Bāṇārasī* Variety in *Khilōnās*

In the previous chapter, we explored the *Bāṇārasī* concept of a *Khilōnā*, and we learned that the *Khilōnā* transcends being a mere playful object for children. It is a cultural signifier shaped by the *Bāṇārasī* worldview. This reflects Alexander's (1964) assertion about the unselfconscious nature of certain design processes, wherein cultures embed a fixed set of myths, traditions, and taboos that are inherently resistant to intentional change and in so doing, limit design variation to what is minimally necessary or imperative. Yet, as this chapter will demonstrate, the *Bāṇārasī* toy designers diverge from this notion in two key aspects: (a) the community does not resist change despite its unconscious or vernacular origin, and (b) the variety of *Khilōnās* made by the community is not limited to functional necessity but panders to desire.

Contrary to the notion of the unselfconscious designer, *Bāṇārasī* toy designers exhibit a keen historical awareness of the shifts in production methods spurred by new technologies and designs. They recognize the necessity for continual evolution to cater to both *Bāṇārasī* patrons and modern customers. Ajit Kumar Viśvakarma, a young toy designer and entrepreneur, exemplifies this adaptability by proposing the use of cloth costumes (as opposed to costumes that are painted on) for wooden *Khilōnās* to streamline production and entice customers with novel offerings. Similarly, shopkeeper Shashi Narayan Agarwal acknowledges the shortcomings of the *Sukhī Parivār* (a nested doll created by *Bāṇārasī* toy designers) in the market, noting its poor finishing, imprecise fitting, and the constant demand for innovation.

This chapter draws upon a manuscript previously published by the author, "*A glimpse inside the Banarasi mind: What do linguistic categories reveal about Banarasi toys*?" in Craft Research (2014), with modifications for clarity and context. The manuscript has been reproduced with permission of the Licensor through PLSclear.

K. Patil, *Toying with the Mind*, Design Science and Innovation,
https://doi.org/10.1007/978-981-96-9227-9_3

Shashi Narayan Agarwal: Different dolls were made here before. They were called *Sukhī Parivār*, or Happy Family. Now, you can see that in the design, all the dolls are of the same kind. But in the *Sukhī Parivār*, there was a man, a woman, and two children, though it lacked a good finish. Now the designs are more finished, so no one looks at the older designs. All these dolls (referring to the saree-clad nested dolls) came later, after seeing the *Sukhī Parivār*. In that set, the man, woman, and children were of different sizes, so they did not fit properly into each other. But in the new dolls, the setting is different. Now you can get any design made—man, animal, and even a god. It is all about painting now.

Author: So, *Sukhī Parivār* went out of production because the dolls did not fit precisely into each other.

Shashi Narayan Agarwal: No, they fit well. But, you see, modern customers want something new. If you buy this today, then tomorrow you will think about what other better features you can get in the same *Khilōnā*.

Transcription 1: Shashi Narayan Agarwal, Agarwal *Khilōnās* Emporium

Interestingly, despite the quest for novelty, *Bāṇārasī* toy designers have managed to create a coherent visual identity that is unmistakably *Bāṇārasī*. Even with the variations among the different types of tiger *Khilōnās* produced in Bāṇāras, a stylistic unity binds them, akin to a family resemblance. Examining the Khilōnās in Fig. 1, one marvels at how, in merely a century, *Bāṇārasī* toy designers, without explicit information transfer, formal organizational structures, or a centralized decision-making body, have achieved such consistency in visual identity, shape, and manufacturing of wooden *Khilōnās*. For instance, while the symbolic meaning of the tiger generally remains constant across *Khilōnās*, the artisanship varies significantly depending on its intended context of use or play.

The tiger in Fig. 1a is carved solid, which reflects the local *Bāṇārasī* belief that religious figurines should not be hollow. Similarly, in the static statue of Goddess Durgā riding her *vāhana* in Fig. 1b, the tiger is also carved solid due to its religious significance. Both these *Khilōnās* are displayed in a *Jhāṅkī*, a tableau created in every *Bāṇārasī* household to depict stories from Lord Kṛṣṇa's life during the festival of *Janmāṣṭamī*.

In contrast, the *Hilantā* tiger in Fig. 1c is constructed by assembling flat wooden cut-outs that incorporate mechanisms for movement and is sold as a playful object for children. Like the carved tigers, this *Khilōnā* is also produced for the *chālū grāhak* (regular local *Bāṇārasī* patrons) and is displayed openly in shops at Vishwanath Gali, *melas*, and *haats* with little variation in features, quality, or price.

On the other hand, the lathe-turned tiger in Fig. 1e on the nested doll embodies a modern sensibility. Its shape is smooth and polished, echoing the Russian nested doll from which it borrows its form and nested structure. The painted features are meticulously delineated. These nested tigers are purchased by a large influx of tourists, including non-*Bāṇārasī* Indians and foreigners. Tourists, especially foreigners, buy

Fig. 1 Variations of *Bāṇārasī* tiger *Khilōnās*. *Image source* Author

(f) (g)

Fig. 1 (continued)

wooden *Khilōnās* they perceive as authentically *Bāṇārasī*. Their expertise in recognizing the *Bāṇārasī* essence of a cultural design may vary, but they prefer *Khilōnās* that depict strong cultural cues and visibly appear hand crafted.

Consider the tiger jigsaw in Fig. 1d as well. Interestingly, after repeated and frequent orders, the jigsaw—a Western game not traditionally played in Bāṇāras—is now being purchased by *Bāṇārasī* patrons—pilgrims and tourists alike. Toy designers have adapted to integrate the Western jigsaw and the Russian nested doll into the *Bāṇārasī* visual identity.

In another instance, the depiction of Goddess Durgā riding her *vāhana*, the tiger, in Fig. 1f significantly differs in body pose from that in Fig. 1b. While both *Khilōnās* are carved, the former is designed to be hung from the ceiling, whereas the latter is designed to be placed on a flat surface. The ceiling-hung *Khilōnā*, called a flying toy, features clearly delineated lines, sharper color contrasts, a stylized body shape, and excessive ornamentation. This *Khilōnā* is specifically produced by *Bāṇārasī* toy designers for modern customers—tourists and foreigners. *Bāṇārasī* patrons, however, do not purchase it because hanging a deity in mid-air is considered inauspicious; a deity should always be kept on a stable, flat surface, as seen in Fig. 1b. Modern customers, unaware of such cultural beliefs, buy the *flying toy* without any concerns about committing sacrilege, unlike the *Bāṇārasī* patrons, who avoid it as a taboo.

The carved tiger in Fig. 1g appears visually similar yet distinct in shape from the other tigers in this group. Its form and pose follow a straight line, unlike the contoured bodies of the other tigers. This Khilona features a round face, stylized stripes, and exaggerated eyes, and resembles a caricature more so than the naturalistic renderings observed in the other *Khilōnās*. This particular *Khilōnā* was a one-time order from a modern client, likely a non-*Bāṇārasī* Indian retailer or exporter. Such modern client orders are a valuable source of livelihood for toy designers, given the declining local consumption of wooden *Khilōnās*. However, the toy designers are unaware of the preferences of these modern customers and merely replicate the designs provided by the clients. The closer the *Khilōnā* matches the client's specifications, the greater the success. Often, *Khilōnās* made exclusively for modern customers bear little or no resemblance to those produced for *Bāṇārasī* patrons. It is noteworthy that *Bāṇārasī* toy designers possess the skill to create *Khilōnās* with a modern sensibility and finish, yet this ability is primarily utilized for modern customers despite its significant impact on the livelihood of the producers.

This simple account of a few tiger *Khilōnās* from Bāṇāras highlights not only the large variety of *Khilōnās* available but also sheds light on the different manufacturing processes, painting techniques, local beliefs, and the heterogeneous consumer population of these *Khilōnās*.

If cultural design thinking is indeed restricted to or limited by patterns of myth, taboo, and tradition, then such variety can only evolve over a considerable period of time. However, literary sources indicate that the *Bāṇārasī* community of toy designers has only come together since the early part of the twentieth century. In barely 125 years, from which close to two-fifths were spent under the colonial yoke, these toy designers have demonstrated remarkable variation in visual representation, shape, and manufacturing. Moreover, despite the extensive variety seen in *Bāṇārasī Khilōnās* over a relatively short period, these *Khilōnās* exhibit a family resemblance—a distinct *Bāṇārasī*-ness or *Bāṇārasīpan*.

The retention of *Bāṇārasī* identity amid this variety suggests the presence of internal mechanisms or underlying organizational principles that support the generation of variations within the boundaries of the *Bāṇārasī* worldview, particularly in the context of *Khilōnā* design. This chapter examines this variation by considering the intuitive and implicit organizational principles used in the *Bāṇārasī* design ecosystem.

2 Assortment Without Possession

Organizational principles profoundly influence thought processes and facilitate recognition, thereby simultaneously streamlining cognitive efforts to comprehend amid extensive variations. We can glimpse into the collective psyche of a community by examining the ways it organizes or categorizes its information—in this instance, *Khilōnās*. Categories not only elucidate for outsiders the connections between *Khilōnās* grouped by a community but also highlight the prototypical instance from which various deviations emerge. The closer a cultural design aligns with its category prototype, the higher its acceptance within the community. Athavankar (1989, 104) posits that the cultural designer's primary intention behind most designs is to conform to the accepted typicality rather than showcase individualism. Cultural designs tend to align closely with their category, embodying typicality through a dialogue between the modern client and the cultural designer. Hence, the closer a cultural design is to the prototypical member of its category, the greater its acceptance within the community.

Rosch and Lloyd (1978, 27–48) assert that typicality is operationally defined by the community's judgment of how well any design fits into a given category.

Crucially, from a design perspective, typicality arising from adherence to precedents remains a benchmark for evaluation and acceptance within such communities.

Distinctions of acceptability and typicality are essential for designing culturally relevant artefacts, especially in a globalized context. However, from a modern perspective, a design strategy relying on typicality and precedents seems like a poor method of creative problem solving and form generation. Alexander (1964, 53–54) claims that when faced with complicated choices, cultural designers will be unsuccessful in inventing forms from scratch. Similarly, Jones (1970, 20) elucidates that in the absence of the drawing board, cultural design innovation is limited to precedents even in contexts that require complete reorganization of the form. Cross (2006, 9) also affirms that many cultural design practices use precedents or previous exemplars. Nevertheless, with rudimentary technology, material-specific skills, low literacy, and absence of foreplaning on the drawing board, cultural designers still produce an astonishing variation in artefacts that often meet the needs of the community (Alexander 1964, 52; Jones 1970, 15).

Therefore, I delve into the question: "How can we account for the design variations produced by a cultural toy designer, despite being creatively constrained by typical instances and traditional precedents?" Accordingly, in this chapter, I explore the kinds and degrees of deviation among typical category members to illuminate design variations within cultural design practices. By understanding artefacts through and between categories—rather than in isolation from their natural groupings—we can decode them as acceptable/unacceptable or typical/untypical, rather than merely distinguishing them as original (in the traditional sense) or new, or as traditional versus innovative. Four key assumptions, underpin this approach.

The first assumption posits that categories can reflect a cultural designer's conception of their world. According to Lakoff (2008, 5–9), organizational principles shape our thinking and aid in the recognition of artefacts; this contributes to cognitive economy amid large variations. We perceive the world not merely as individual entities but also as categories of entities. Indeed, we conceptualize, organize, and recognize our world through these categories. Thus, categories not only reflect our mental concepts but also our understanding of the world, particularly in the context of artefacts.

The second assumption is that categories encompass both typical and deviant instances. Categories inform an outsider about the linkages between artefacts grouped by a community and reveal the prototypical instance of each group, from which varied instances can deviate to different extents. Hence, a typical instance of a cultural design category provides a benchmark from which possible deviations in cultural design can occur.

The third assumption contends that precedents shape cultural design thinking. The close fit between form and context in most cultural designs arises from a continuous process of iteration and improvization of precedents (Alexander 1964, 37–48; Jones 1970, 20). John further considers precedents as an essential store of information akin to genetic coding. However, precedents here are not random or arbitrarily chosen instances that simply precede the variation chronologically. More importantly, precedents share properties of time and kind within the same category or across multiple categories. A shared set of properties may include techniques, identity, and functions, among others, and may thereby influence the perception of cultural design itself.

The fourth assumption is that categories reflect design variation or the lack thereof in cultural designs. In his book *Artefacts as Categories*, Miller (1985) distinguishes 31 different varieties of pots produced by potters in Dangwara, India, along with their uses.[1] The variety revealed in the categorization of Dangwara pots not only reflected the potters' ability to conceive of and produce variations but also the complex patterns of food preparation, eating habits, and the rituals of every meal within the community. The *Bāṇārasī Khilōnās* have similarly exhibited such variations.

These observations suggest that studying the categorization of design can shed light on a cultural designer's decisions that shape the acceptability and novelty of a cultural artefact. Therefore, this chapter focuses on the community-held categorization of the *Bāṇārasī Khilōnās*. Here, *community* refers to the stakeholders of the *Bāṇārasī Khilōnās*: primarily the *Bāṇārasī* toy designers who conceptualize and produce the *Khilōnās*, the *Bāṇārasī* patrons and modern customers whose tastes influence the *Khilōnās*, and the shopkeepers in the Vishwanath Gali who mediate between the artisanship of the toy designers and the preferences of the patrons and customers. Coomaraswamy (1909) astutely articulates the relationship shared by the stakeholders in a community:

It is the business of the artist to know how things ought to be made and to be able accordingly, as it is the business of the patron to know what things ought to be made, and of the consumer to know what things have been well and truly made and to be able to use after their kind.

3 Are *Mūrti* and *Khilōnā* the Same Object?

Mūrtis (statues of gods) are not called *Khilōnās*. In fear of the children and also because they belong to god, these *Khilōnās* are carefully packed and kept aside for the next year. There is always a desire to own more *Khilōnās*, so every year some new ones are also purchased.

Transcription 2: Suman Shrivastava, a housewife residing in Laksha, Bāṇāras

Suman Shrivastava, a housewife, shared her thoughts on the *Khilōnās* she and her family used to decorate the *Jhāṅkī* during *Janmāṣṭamī*. Interestingly, she distinguishes between a *Khilōnā* and a statue of a god. She refers to the statue of a god as a *Mūrti*, which she handles with care and protectiveness. However, in the same

[1] Miller documented nearly 31 names for different kinds of pots and their uses as produced by potters in Dangwara, India. Such variety is virtually unheard of in modern kitchen utensils. The diversity of pots in Dangwara not only showcases the potters' ability to conceptualize and produce variations but also mirrors the intricate patterns of food preparation, eating, storage, and community rituals. The purposes of these pots are intertwined with their cultural design, as reflected in the 31 names used in the language of the Dangwara community.

conversation, she also refers to the *Mūrti* as a *Khilōnā*. This raises intriguing questions: How can a *Mūrti* be a *Khilōnā* and not a *Khilōnā* at the same time? Are *Mūrti* and *Khilōnā* the same objects? Or is *Khilōnā* a broader category that includes *Mūrtis* as sacred objects alongside other nonsacred objects given to children?

This distinction in vocabulary prompts further inquiry into whether the lexicon, as discussed in the previous chapter, reflects the variety of Khilōnās. Can the *Bāṇārasī* lexicon account for these variations in the same way that Suman Shrivastava distinguishes between a *Mūrti* and a *Khilōnā*? Among the tiger *Khilōnās* discussed in the preceding section, how can we determine whether Fig. 1a is a *Mūrti* and 1c is a *Khilōnā* intended for playful purposes?

The authenticity of cultural design thinking can only be revealed if the captured categorization of the *Khilōnās* genuinely belongs to the *Bāṇārasī* community itself and is not imposed by me or any other researcher as an outside observer. In this regard, the vernacular language could be a valuable tool for uncovering the community's organizational principles and thereby avoiding researcher bias. Using vernacular vocabulary allows for forming an insider's view of the community's world, one that is less influenced by the researcher's own classification system.

The lexicon of a culture reflects how people perceive their environment and what is most important to them (Serena and Warms 2007). Athavankar (1989) also concurs that natural language can manifest a category system and influence how people structure their world. Both typical and deviant members of a category can be represented in natural language. Once a term is accepted into the language, it also impacts how the cultural design is perceived within the culture.

To gain authentic insight into the principles embedded by cultural designers in their work, it is imperative to capture the categorization of artefacts as used by the toy-design community rather than impose an external perspective. Vernacular language serves as a valuable tool for elucidating a community's organizational principles because it reflects the local or insider view of the community's world without the researcher's own classification bias (Serena and Warms 2007, 116–135). Additionally, typical and deviant members of a category can manifest in language (Athavankar 1989, 101; Krippendorf 2006, 147). Therefore, I will attempt to capture the community-held categorization of the *Bāṇārasī Khilōnās* through the *Bāṇārasī* lexicon itself.

4 How Do Toy Designers Speak of *Khilōnās*?

Let us examine extracts from transcripts of toy designers, consumers, and shopkeepers for cues that reflect the community's perception of *Khilōnās*. According to Krippendorff (2006), nouns evoke ideal types of categories that enable competent speakers of a language to distinguish among *Khilōnās* based on their typicality within available categories. Therefore, proper nouns and collective nouns used repeatedly to denote *Khilōnās* have been observed in these extracts. The characteristics describing the nouns were captured through attributive adjectives or prepositions. In the extracts discussed below, nouns are underlined, while adjectives or prepositions describing the nouns are highlighted in grey.

4.1 Toy Designers

In *fancy work*, the eyes, nose, and clothes are *Gadhwā* (carved by hand). It looks ***beautiful*** and is ***well-paid*** for. ***Outsiders*** buy it.

Local work is ***small*** in size. The *figure* is represented by a painting on *wood*.

Transcription 3: Ramsingh Maurya, toy designer

Saras (Heron) is a *plain Khilōnā* (unpainted artefact or one with flat monochromatic painting). It is a *Pattern kā Khilōnā*, that is, it is cut on a ***flat wooden plank*** and ***assembled***. *Decorative Khilōnās* are also *plain Khilōnā*.

Transcription 4: Sunder Viśvakarma, toy designer

A *Khilōnā* made of *Sāgavān* wood is ***fancy***, whereas one made of *Gūlar wood* is ***local***. ***Sada gend*** (ordinary ball) is made of *Gular*, and ***fancy*** *Shataranj* (chess) is made in **expensive** wood.

Transcription 5: Shyam Viśvakarma, toy designer

The *Mūrti* (sacred statue) of *Rādhā* (consort of God Kṛṣṇa) holds a *lotus* in her hand and *Gopī* (milkmaid) holds a *pot* in her hand.

Transcription 6: Ratan Prajāpati, toy designer

The lexicon employed by *Bāṇārasī* toy designers unveils a profound tapestry of terms that encapsulate not only the essence of a *Khilōnā* but also the intricate processes of its creation, its symbolic representation, and its multifaceted functions. This shared understanding is deeply embedded in the lexicon of toy designers, which prominently features terms related to the manufacturing process, the representation of the *Khilōnā*, and the meaning and purpose associated with its consumption.

The production process encapsulates various skill sets possessed by the toy designers. These include creating a *Khilōnā* using machinery, referred to as *Kharād*; hand-carving, known as *Gadhwā*; and assembly techniques like *pattern kā Khilōnā*. The production vocabulary is further enriched by nouns describing materials, such as different types of wood (*Sāgavān*, *Gūlar*) and paint (*Babūl* glue). These diverse skill sets are not only evident in the lexicon but also observed in the *Karkhanās*—small, specialized workshops that function independently under different roofs, reflecting a modern-day segmentation of factory floors.

Toy designers are acutely aware of consumption patterns, and they recognize that *Khilōnās* serve various purposes, including as decorative items or religious idols (*Mūrti*). This awareness is heightened by their own consumption of *Mūrtis*, which they often use to create *Jhāṅkī* at home. Consequently, they can vividly illustrate the functional context and symbolic meaning of these *Khilōnās*. For example, many toy designers are familiar with the story of the popular parrot *Khilōnā*, identified as Sukhdev, the parrot god.

In addition to production and function, the lexicon of toy designers heavily features terms that describe the representation of a *Khilōnā*, including labels like Plain, Sada, *Fancy*, and Local. Notably, some *Khilōnās* derive their names directly *from their visual attributes, such as Fancy Gend (fancy ball) or Sada Shatranj* (plain chess). It is intriguing to note that many of these descriptive terms are borrowed from English rather than derived from vernacular languages. Representation serves as the externalization of cultural design ideas, and is often the first point of contact between the product and its potential buyer, and can influence purchasing decisions. Therefore, the lexicon of representation not only reflects the toy designers understanding of patron preferences but also the preferences themselves.

This bidirectional influence suggests that changes in consumer vocabulary will also affect the lexicon of representation used by toy designers. Additionally, many foreign customers communicate exclusively in English and explain their choices and preferences through adjectives such as *plain.* This interaction further embeds English descriptors into the local lexicon and demonstrates the dynamic nature of language in the context of cultural production and consumption.

This shared vocabulary facilitates communication among toy designers with shopkeepers and occasionally with outsiders like myself. Table 1 presents a detailed analysis of the lexicon used by a sample of ten toy designers and offers insight into the rich linguistic landscape that underpins this cultural design community.

Table 1 Sample analysis of the lexicon of toy designers

No.	Part of speech	Semantic
1	Proper noun referring to an object	*Saras, Hilantā, Rādhā-Kṛṣṇa, Flying toy, Sukhī Parivār, Dānevālī chiṛiyā, Butler*
	Adjective referring to representation of the proper noun of an object	*Fancy, Plain, Local*
	Adjective referring to meaning of the proper noun of an object	*Rādhā* holds a lotus in her hand
2	Proper noun referring to a material	*Gūlar, Sāgavān, Babūl* glue
	Adjective referring to an object's property	*Babūl* glue is strong
3	Noun referring to the modern customer	*Decoration, poojaniya* (sacred)
	Adjective referring to use	High price
	Adjective referring to value	
4	Proper noun referring to a production process	*Gadhwā* (carving)
	Verb referring to a production process	*Katana* (cutting), *Kharād* (lathe turning), *Khudāi* (digging)

4.2 Bāṇārasī *Patrons*

God's figurine is not considered a *Khilōnā* but a *Mūrti*.

Transcription 7: Aarti Shrivastava, *Bāṇārasī* resident

Children play with it therefore they are called *Khilōnās*.

Transcription 8: Parvati Gupta, *Bāṇārasī* resident

Mūrtis are for ***decoration***, but these *Khilōnās* are treated with ***respect***. We take ***care*** not to touch them with our *feet* or that *children* don't kick them around. However, *plastic toys* are given to *children*.

Transcription 9: Krishna Sharma, *Bāṇārasī* resident

All *Khilōnās* are offered to *God*. The ***broken ones*** are given to *children*.

Transcription 10: Archana Agarwal, *Bāṇārasī* resident

In comparison to the toy designers and shopkeepers, the lexicon employed by the *Bāṇārasī* patrons is notably restricted to notions of use, meaning, and personal preference rather than extending to matters of production, representation, or contemporary market segments. However, the patrons remain a crucial component of the *Bāṇārasī* community of toy designers, as the patrons' purchasing power underpins the cultural design and, consequently, the livelihood of the toy designers. *Bāṇārasī* patrons select and utilize *Khilōnās* that resonate with the shared worldview of both producers and users. The intended meaning or function of the *Khilōnā* determines its appropriateness within this collective worldview.

Thus, when distinguishing the noun *Mūrti* from *Khilōnā*, *Bāṇārasī* patrons emphasize the distinction based on meaning or function. This distinction is not only articulated through the specific noun but also through attributive propositions that specify the user, the context of use, preferences, and cultural references, as illustrated in Table 2. The patrons are well versed in the narratives that many *Mūrtis* embody and often articulate the role these objects play in the cultural and moral development of children.

Table 2 Sample analysis of the lexicon of consumers

S. No.	Part of speech	Semantic
1	Attributive proposition of use	Handle with care
2	Noun denoting the user	Not given to children
3	Noun denoting the space of use	*Jhāṅkī*
4	Proper noun of occasion of use	*Janmāṣṭamī*, *Shaadi*, *Ramleela*, *Diwali*
5	Adjective of preference	*Asābhyā* (uncouth)
6	Noun denoting cultural reference of stories, events, and activities	*Vrat* (fast), *Nāga-nāthīya*, *Kārāgara* (jail)

The lexicon encompassing the meaning and function of the *Khilōnā* is a common ground shared among toy designers, shopkeepers, and customers alike. However, it is intriguing to observe that *Bāṇārasī* patrons are unfamiliar with the collective term *Decoration kā Khilōnā* despite its frequent use among *Bāṇārasī* shopkeepers and toy designers. This discrepancy may arise from the fact that *Bāṇārasī* patrons are not the primary consumers of Decoration *kā Khilōnās*; these items are predominantly purchased by modern customers. Consequently, as these decorative *Khilōnās* are not part of the generational *Bāṇārasī* usage, the terminology does not permeate the local consumer lexicon.

4.3 Bāṇārasī *Shopkeepers*

This is the *King*. You can keep this in *Jhāṅkī* or for ***home decoration***. In *Bāṇāras*, you can also use it as *a gift*. Nothing more, this is not a ***sacred*** *Khilōnā*.

Transcription 11: Nitin Saxena, Shopkeeper

Clients from ***outside*** *Bāṇāras* ask for *Śaṅkara* (Hindu God). It has no importance in any *festival* but it is made nevertheless to meet the demand.

Transcription 12: Behari Agarwal, Shopkeeper

There were many *toy designers* who made ***Kalkattiya*** *boxes* before. They made ***Plain*** *boxes without flowers or petals*. They used ***English colours*** to paint it. There was ***a lot of space*** inside the box. The boxes were a ***favorite*** amongst the ***Bengali*** *consumers*. (Bengal is a state in Eastern India.)

Transcription 13: Shyamlal Agarwal, Shopkeeper

The shopkeepers serve as intermediaries in the transaction between producers and consumers, and as such they share the vocabulary of both the toy designers and the patrons. In the Vishwanath Gali, shopkeepers maintain close bonds with toy designers. The exclusive market for wooden *Khilōnās* developed alongside its cultural design, thus creating parallel lineages of shopkeepers and toy designers. For instance, the father of Ratanlal Prajāpati, a toy designer, sold *Khilōnās* to the father of Beharilal Agarwal, a shopkeeper. Today, their sons continue this relationship.

Beyond business transactions, shopkeepers sometimes assist toy designers in making informed cultural and design decisions based on market demand for modern customers, to which they are more attuned. During a visit to the Agarwal toy shop, I heard Beharilal Agarwal placing an order to a toy designer. In addition to specifying the number of pieces and sizes required, he repeatedly emphasized that "Yeh bahar kā kām hai" (which means, "This is an outside order"). Hence, local work would not be accepted, and good *fancy* work was needed.

Like toy designers, shopkeepers possess a rich vocabulary related to the representation, production, and meaning of *Khilōnās*. This knowledge is crucial for interpreting and explaining modern client orders and preferences to toy designers. A shared understanding of *Khilōnās* and their artisanship, expressed in vernacular language, facilitates the interpretation of client needs and the delineation of work orders among toy designers. The shopkeeper's role in designing a *Khilōnā* can vary from outlining a design need to explaining or elaborating on it in the modern client context and even supporting ideation in some cases; all of these roles are facilitated by a shared lexicon with the toy designer.

Outside the *Karkhanā*, shopkeepers are the first filter of acceptance or rejection of a *Khilōnā.* If a shopkeeper deems a design inappropriate for modern customer preferences, he can send it back to the *Karkhanā* for iterations.

In addition to production and representation, the lexicon of the shopkeepers includes specific references to various modern customer segments such as *chālū grāhak*, Bengali *grāhak*, tourists, and foreigners. These terms indicate the type of customer likely to purchase a *Khilōnā* based on its function or use—be it for decoration, as a *Mūrti*, or as a *Khilōnā* for play (*Khelne kā Khilōnā*). This specialized vocabulary reflects the modern customer segments and is unique to shopkeepers, who observe consumption patterns more closely than toy designers (who are often confined to their *Karkhanās*).

Toy designers, patrons, and shopkeepers collectively construct the *Bāṇārasī Khilōnā* through their use of specific nouns, verbs, and adjectives. These terms are not isolated to individuals but are repeatedly used by various community members, as seen in the transcriptions presented in this chapter. The shared language used to describe a *Khilōnā* suggests a community-wide lexicon that varies in its extent of use among different members.

The vernacular lexicon discussed here, along with the list of *Khilōnās* from the *Varanāsi Kaṣṭha Khilōnā Samiti* (Varanasi Society for the Purchase of Wooden *Khilōnās*), strongly indicates the possible existence of a community method for categorizing and defining *Khilōnās.* A small study was undertaken to triangulate these findings with the *Bāṇārasī* community lexicon and the nomenclature used in the *Varanāsi Kaṣṭha Khilōnā Samiti.*

5 Cultural Design Sorting: A Thought Elicitation Tool

At this juncture, it is essential to articulate certain observations about toy designers as sources of information in this study. As an outsider to the Bāṇārasī community, my interactions with the toy designers were marked by a degree of distance. Despite their willingness to discuss and recount their histories and work processes, their manner of understanding and interpreting situations differs from mine due to our divergent worldviews. Consequently, even when toy designers articulate a notion, it is not always possible for me to fully apprehend the situation from their perspective.

This disparity underscored the need to identify methods that would allow toy designers to directly demonstrate or explain their decisions regarding a *Khilōnā.* Given their low levels of literacy and limited or nonexistent familiarity with contemporary technological devices, I explored the possibility of initiating dialogue through real *Khilōnās* or their images to create a context for discussion. Toy designers are most comfortable with the physicality of the *Khilōnās*, so it was presumed that they would be able to elucidate their decisions more effectively through discussions or activities involving these tangible objects.

Given that the use of physical *Khilōnās* limited the number of instances that could be discussed and compared at a given time, images of the *Khilōnās* were used to facilitate the sorting process. Sorting images does not necessarily require verbal elicitation, although such verbal input can enhance understanding of the decisions made during the grouping process. Sorting can be carried out independently by individuals, so it minimizes the potential influence of the researcher on the study. This method allows for multiple selection attempts and decision-making processes with minimal resources and time gaps between iterations. The iterative attempts, confusions, and delays between decisions can provide insights into the levels of difficulty, familiarity, or confusion experienced during the decision-making process.

To explore this method, a mixed assortment of 88 images of typical and atypical *Khilōnās* from various *Bāṇārasī* categories, as well as non-*Bāṇārasī Khilōnās*, were presented to two toy designers, two shopkeepers, and two patrons. Each card featured a single image of a *Khilōnā*. Participants were asked to sort the cards into groups of *Khilōnās* they thought were similar to each other. This exercise was unfamiliar and abstract for the participants, so I built a context around the activity to make it meaningful for them.

The participants were informed that I was designing a Cultural Design Museum,[2] which would include a section on *Bāṇārasī Khilōnās*. However, I did not know how to arrange the objects on the shelves since some *Khilōnās* are sacred, others playful, and yet others modern and abstract. The participants were asked to advise me on the best way to display the *Khilōnās* on the shelves. There was no limit on the number of groupings they could create, and they were allowed to change the grouping at any point. This method of card sorting not only facilitated categorization but also created a context that encouraged the participants, especially the toy designers, to be more vocal in sharing their stories and descriptions. The images of the *Khilōnās* are shown in Figs. 2, 3, 4, 5 and 6.

For the card sorting exercise, 30 images of *Mūrtis* designed with varying levels of skill, both old and new, as seen in Fig. 2, were selected. Some were used in *Jhāṅkīs*. All *Mūrtis*, whether depicting deities or common folk, were intricately carved, while some animal figures were *pattern*-cut and featured round faceting to simulate hand carving. Among these, eight statues on the far right are currently sold as antiques and thus correctly fall into the Decoration *kā Khilōnā* category. However, the question arises: Will this new grouping be recognized by the community? The card sorting exercise also helps reveal the community's perception of the *Mūrtis*' historical and

[2] Such a museum was indeed designed, called- Śilpālay at the Indian Institute of Technology in Kanpur, India with funds received from the Government of India. Besides other handmade designs from clusters such as Morādābād, Saharanpur, Gorakhpur, etc., the museum also has several works from the *Bāṇārasī* community of toy designers.

Fig. 2 Cultural design sorting cards of *Mūrtis*. *Image source* Author

cultural significance and how these perceptions align with the functional categories used in contemporary market contexts.

For the card sorting exercise, 27 *Khilōnās* intended for ludic (playful) purposes were also included (refer to Fig. 3). All these *Khilōnās* were machine made, either on a lathe or a cutter, with one exception that was carved. Notably, five *Khilōnās* on the far right were made in *Bāṇāras* but exclusively for modern client orders, and therefore they are not in regular production.

In addition to the previously mentioned categories, 22 *Khilōnās* that serve neither a religious nor a ludic purpose were included in the card sorting exercise (refer to Fig. 4). These *Khilōnās* were produced either by hand or machine, and both old and new unpainted *Khilōnās* were mixed with the rest of the images. Among these, four *Khilōnās* on the far right were created for one-time modern client orders and are no longer in regular production in *Bāṇāras*.

To further verify the understanding of the community, four additional *Khilōnās* made by other cultural design communities of *Bāṇāras* were included in the card sorting exercise: one old homemade *Bāṇārasī* doll made of cloth and three papier-mâché masks made in *Bāṇāras* for Ramleela (refer to Fig. 5). These items have familiar meanings but differ in material, representation, and manufacturing. The objective is to see if the *Bāṇārasī* community recognizes these *Khilōnās* despite the

Fig. 3 Cultural design sorting cards of *Khelne kā Khilōnās*. *Image source* Author

differences and to understand if they will categorize them in the same way as wooden *Khilōnās* or differently.

In the assemblage of images, the deviant cultural designs are embodied in instances of modern toys (refer to Fig. 6). Among these are two wooden pecking toys that were designed using conventional techniques such as lathe turning and *pattern* assembly. While their material and mechanisms resonate with familiar practices, their representations diverge from generational norms. The set included a plastic Barbie doll, which was a stark departure with its foreign manufacturing process, symbolic significance, and synthetic composition. Besides these, there were also two wooden toys whose fabrication employed recognized methods and materials but whose forms and meanings remained alien to the *Bāṇārasī* toy designers.

The pertinent inquiry here concerns how the *Bāṇārasī Khilōnā* community will engage with these non-native toys. Will they be able to identify and relate to these modern designs? And how will these contemporary artefacts be positioned in relation to the time-honored *Bāṇārasī Khilōnās*? This exploration seeks to unravel the community's capacity for recognition and integration of these divergent forms within their established cultural frameworks.

For brevity, the analysis of the card sorting by one toy designer, *Baicchan Singh Maurya*, is discussed here. During this exercise, a young apprentice accompanied the

Fig. 4 Cultural design sorting cards of *Decoration kā Khilōnās*. *Image source* Author

Fig. 5 Cultural design sorting cards of *Khilōnās* made by other cultural design communities in Bāṇāras. *Image source* Author

(a) (b)

Fig. 6 Cultural design sorting cards of non-*Bāṇārasī Khilōnās*. *Image source* Author

Fig. 7 Sorting by *Baicchan Singh Maurya*: *Mūrti*. *Image source* Author

subject and offered suggestions on the grouping. *Baicchan Singh Maurya* categorized the cultural design images into four distinct groups:

1. ***Mūrtis*** (Fig. 7)
2. ***Bacchon ke khelne kā Khilōnā*** (*Khilōnās* to be played with by children) (Fig. 8)
3. ***Sajāvat (Decoration) kā Khilōnā*** (Fig. 9)
4. Those he could not comprehend and therefore discarded (Fig. 10).

Most *Khilōnās* that the subject grouped together as *Mūrtis* held religious significance (refer to Fig. 7 and Table 3). These items were all carved and painted. Interestingly, the subject also included two old, unpainted carved *Mūrtis* in this group and distinguished them from *Khilōnās* meant for placement on *Almārī* or for decoration. This distinction indicates an awareness of the recent convention of using unpainted carved figurines of deities solely for decoration in contrast to the traditional practice. The toy designer recognized this distinction and thus showed an understanding of evolving practices.

Additionally, the subject assigned multiple functions to figures of animals and birds and recognized their roles in both display within a *Jhāṅkī* and in ludic play. This dual function suggests that ludic play with *Khilōnās* of religious value also has an enculturative aspect that integrates religious significance with play.

Fig. 8 Sorting by *Baicchan Singh Maurya*: *Bacchon ke khelne ke khilōne*. *Image source* Author

Fig. 9 Sorting by *Baicchan Singh Maurya*: *Decoration kā Khilōnā*. *Image source* Author

Fig. 10 Sorting by *Baicchan Singh Maurya*: Discarded images. *Image source* Author

The subject grouped all secular *Khilōnās* together and designated them as *Khilōnās* to be played with by children (refer to Fig. 8 and Table 4). This category included lathe-turned and pattern-assembled *Khilōnās*, most of which exhibited some form of movement. While organizing these *Khilōnās*, the subject frequently mentioned and reiterated the primary users of the *Khilōnās*—children.

One notable exception in this group was a *plain* pecking *Khilōnā* that the subject did not recognize as a non-*Bāṇārasī Khilōnā*. Given that pecking toys are also commonly made in *Bāṇāras*, it is plausible that the subject mistakenly identified this non-*Bāṇārasī* pecking toy as a local product. This instance highlights the possibility for overlap and misidentification when cultural forms are shared across different contexts or in modern spaces without a context.

The subject independently introduced the concept of spaces of display for *Khilōnās* and mentioned locations such as the *almārī* (locker) and the *mehmān ghar* (drawing room) (refer to Table 5). In rural India, the flat top of a locker is commonly used to display showpieces. The subject applied this familiar context in discussing *Decoration* or *Sajāvat kā Khilōnās* and often used the Hindi term *Sajāvat* interchangeably with *Decoration*. During the sorting process, the subject frequently altered the groupings of Khilōnās with religious connotations. For instance, he moved an unpainted Kali figure from the *Mūrti* group to the *Decoration kā Khilōnā* group.

Similarly, a *flying toy* was shifted from the *Khelne kā Khilōnā* (*Khilōnā* for playing) group to the *Decoration kā Khilōnā* group. Conventionally, *Khilōnās* with religious meanings were classified as *Mūrtis*. However, with the rise of non-*Bāṇārasī* consumption, religious figurines are increasingly sold for display rather than for

Table 3 Cultural design card sorting of *Mūrtis* by *Baicchan Singh Maurya*

S. No.	Toy designer: *Baicchan Singh Maurya, Gadhwā Khilōnā*, Khojwa, Bāṇāras		
	Field trip held in January 2012		
	Image	Transcription translation	Remark
1		Toy designer: This is a *Sakhī*. She is playing a *Baja* musical instrument. She and other such toys are used for decoration. They are kept close to the idol of the deity	The toy designer is aware of the meaning and use of the *Khilōnā* as shaped by the worldview
2		Toy designer: You can keep the image of a god in the temple. You can also keep this next to the god	These are birds used in the *Jhāṅkī* during *Janmshtmi*
3		Toy designer: This is can be placed along with the goddess because a cow is holy. It can also be used for display The cow is also kept with Kṛṣṇa. You can also give it children. Children will play with mother cow. So you can display it, give it to a child, or keep it on the almari, f you please	The toy designer notes the holy status of the cow but still thinks that children can play with it. But in the child's play he takes care to mention mother cow. This is similar to the notion of enculturative play mentioned in Chap. 2 This is an antique piece, so the toy designer also states that it can be kept on the locker for decoration. It is interesting to note that the cultural design belongs to multiple groups at the same time

Image source Author

ritual or ceremonial use (refer to Fig. 9). This hesitation and confusion in categorizing *Khilōnās* of a religious nature underscores the strong influence of cultural worldviews on the subject's decision-making process.

All *Khilōnās* that were unfamiliar or unfit within the *Bāṇārasī* worldview were set aside by the subject (refer to Fig. 10). He correctly recognized all the non-*Bāṇārasī* toys but was unable to label the separate group or provide a rationale for their exclusion. He claimed to have no understanding of them. One cultural design was termed "scary" and deemed unfit to be kept in a room or given to a child to play with (refer to Table 6, row 2).

Baichan Singh Maurya, the toy designer performing this card sorting, was able to group *Khilōnās* as of a religious nature, as used for ludic play, and as used for

Table 4 Cultural design card sorting of *Bacchon Khelne Kā Khilōnās* by *Baicchan Singh Maurya*

S. No.	Toy designer: *Baicchan Singh Maurya, Gadhwā Khilōnā*, Khojwa, Bāṇāras		
	Field trip held in January 2012		
	Image	Translation	Remark
1		Toy designer: These are for children to play with	
2		Toy designer: These are for children to play with	This *Plain* lathe-turned *Khilōnā* is not made in Bāṇāras. The pecking *Khilōnā* made in Bāṇāras is a painted *pattern kā Khilōnā*
3		Toy designer: These are for children to play with	

Image source Author

display, along with their meaning, user, and place or context of use. He also provided additional information about the manufacturing of the *Khilōnās* (refer to Table 7). Some *Khilōnās* were not placed in any specific manufacturing group and were simply called machine made. These *Khilōnās* are designed using a newly acquired skill of faceting a solid piece of wood on a fret saw to simulate the effect of carving. This skill does not fit into the cultural categories of Kharād, Patri, or Khudāi, and the *Bāṇārasī* lexicon has not yet acquired a new word for it. As a result, the subject did not include them in any specific manufacturing group (refer to Table 7, row 6).

A similar activity was carried out with one more toy designer, two consumers, and two shopkeepers. The results were largely similar but with a few notable exceptions. The toy designer and the shopkeepers sorted the older *Mūrtis* differently. The older *Mūrtis* of Gaṇeśa, Viṣṇu, Lāḍū Gopāl, and so on were traditionally used in *Jhāṅkīs* (refer to Fig. 11). However, as time has passed, these larger *Mūrtis* have entered the market for sale at high prices and are often bought by connoisseurs as antiques. Consequently, the shopkeepers tend to group them together under *Decoration*. In contrast, the toy designers still group them with *Mūrtis* for display in *Jhāṅkīs*. This difference is natural because toy designers are unaware of the manner in which these old *Khilōnās* are now being consumed in the market.

Most toy designers could not understand or comprehend abstract *Khilōnās* and, therefore, simply discarded them, as seen in Fig. 10. However, some of these discarded *Khilōnās* were recognized by the shopkeepers as goods for export, which they indeed are. Given their familiarity with the export market and its preferences,

Table 5 Cultural design card sorting of *Decoration/Sajāvat Kā Khilōnās* by Baicchan Singh Maurya

S. No.	Toy designer: Baicchan Singh Maurya, *Gadhwā Khilōnā*, Khojwa, Bāṇāras		
	Field trip held in January 2012		
	Image	Translation	Remark
1		You can also place this on the locker	The flat top of the locker is often used as an area for display in Indian village households. The toy designer uses the analogy of the locker very often to indicate the decorative status of the cultural design. In this case, despite the image of a god painted on the nested doll, the *Khilōnā* is not considered sacred
2		Toy designer: This is for a child to play with Author: But isn't she a goddess? Toy designer: Yes, she is sitting on the parrot. She is a goddess Author: So can a goddess be given to a child to play with Toy designer: No. We won't. She is to be hanged Author: Then can the hanging one be given to a child? Toy designer: No, why should it be given to a child? She will be hanged in the room. Otherwise, where will we keep her?	Note the change in the grouping as soon as the toy designer realizes that she is a goddess; nonetheless, the cultural design was not placed along with the other sacred *Khilōnās*. The toy designer may not have been able to recognise this *Khilōnā* because it is usually carved but in this instance is made with a flat *pattern* of wood assembled together

(continued)

Table 5 (continued)

S. No.	Toy designer: Baicchan Singh Maurya, *Gadhwā Khilōnā*, Khojwa, Bāṇāras		
	Field trip held in January 2012		
	Image	Translation	Remark
3	 *Bāṇārasī Khilōnā*	Toy designer: No, change this. These are *Mūrtis* to be decorated in the temple Author: This one too? Toy designer: Yes she is *Kālī* Author: Is *Kali* put in *Janmashtmi* or in any other festival without being colored? Toy designer: Yes she is without color. She is *Plain*. Yes she can be kept. It depends on you. Wherever your mind goes, you can keep it. But if we think of where to keep her, then maybe on the locker. If someone comes they will see her. Like that you can keep here wherever you want	The toy designer has made a distinction between the way he thinks I would/could use this cultural design and the way he would. He presumes that I have no knowledge of religious *Khilōnās*, so he may keep even an un-colored *Plain Mūrti* in the temple. On the other hand, he himself will use it on the locker for display so that others could see but would not keep it in his temple He clearly states that I can put the cultural design wherever I wish, thus relieving me of the conventions and customs of his own worldview
4		Toy designer: This will be placed on the almari in the drawing room. Guests will come and see it. Anyone can see it	The toy designer is either not familiar with or cannot understand that the cultural design is a puzzle. He places it again on the almari for display and mentions the space in which the locker is kept: the drawing room (emphasizing the purpose of "show")
5	 *Bāṇārasī Khilōnā*	Toy designer: This is a crane. You can also keep this on the locker	
6	 *Bāṇārasī Khilōnā*	Toy designer: This is a king riding his elephant. You can keep it on the locker	

(continued)

Table 5 (continued)

S. No.	Toy designer: Baicchan Singh Maurya, *Gadhwā Khilōnā*, Khojwa, Bāṇāras		
	Field trip held in January 2012		
	Image	Translation	Remark
7	 	Toy designer: These are all made with machines; they are *Kharād kā kām*	

Image source Author

Table 6 Cultural design card sorting of discarded images by Baicchan Singh Maurya

S. No.	Toy designer: *Baicchan Singh Maurya*, *Gadhwā Khilōnā*, Khojwa, Bāṇāras		
	Field trip held in January 2012		
	Image	Translation	Remark
1		A: Is this our *Khilōnā* from Bāṇāras? C2: This is a *Kharādi* work A: I have seen it being made in Bāṇāras C1: This is made in Bāṇāras but I can't understand it. Take it out from these	
2		What is this? Look at this, I can't understand this. Some image … fearsome image. No one keeps a scary object in a room He discards the image	The shape of this *Khilōnā* is unfamiliar to the toy designer even though its visual representation is similar to that of the *Bāṇārasī* painting. But due to the unfamiliarity of the shape, he considers the *Khilōnā* to be scary. The worldview also does not permit scary objects to be kept in the room

Image source Author

Table 7 Names and descriptions of various groups formed by *Baicchan Singh Maurya*

S. No.	Type of *Khilōnā*	Instances of *Khilōnās*
1	*Local kām* It is cheap and of low quality. Less work done	
2	*Fancy Kām* All *Khudāi* work will be generally costly and *Fancy*. It depends on painting as well as shaping. All work done on machine is *Fancy*	
3	*Plain Kām* Unpainted cultural design	
4	*Kharād kā kām* Lathe-turned *Khilōnās*	
5	*Khudāi kā kām* Carving *kā kām* Wood work which is hand carved with different parts joined together. *Mūrtis* are carved and assembled, but they cannot be worshipped because the joints desecrate them according to the Hindu religion	
6	Machine *kā kām* Not defined; but subject mentions that it is cut on a template	
7	*Patre kā kām* A *gaddi* (a round peg) is sliced into flat plates (*Patri*). The plates are cut into various parts of the design, which are assembled to make a *Khilōnā*. It is not through shape but painting that you can recognise this *Khilōnā*	

Image source Author

Fig. 11 Old *Mūrtis* used in *Jhāṅkī* earlier but now sold as antiques. *Image source* Author

shopkeepers were better equipped to recognize these unfamiliar *Khilōnās* (refer to Fig. 12).

As discussed with the lexicon observed in the transcriptions, *Bāṇārasī* patrons share only a limited vocabulary with the toy designers and shopkeepers regarding the *Khilōnās*. This vocabulary is largely restricted to the meaning and purpose of the *Khilōnās* and the myths, stories, rituals, and ceremonies woven around them. It was observed again in the card sorting activity that *Bāṇārasī* patrons are unfamiliar with the collective noun *Decoration kā Khilōnā*, a term heavily used by both *Bāṇārasī* shopkeepers and toy designers, as noted earlier. Since the *Bāṇārasī* population is not the primary patron of *Decoration kā Khilōnās*, these *Khilōnās* and their nomenclature are not part of their lexicon.

The next section explores the lexicon and organization of *Bāṇārasī Khilōnās* as understood by the community and traces the evolution of these elements over time.

Fig. 12 *Khilōnās* of export familiar to shopkeepers. *Image source* Author

6 The Category Cube

The vernacular lexicon used by the toy designers, patrons, and shopkeepers of Bāṇāras reveals multiple ways of organizing *Khilōnās* and demonstrates that a single *Khilōnā* can belong to more than one group. Some organizational principles are shared among all community members, while others are limited to a select few. This analysis discusses the three basic groups of organization identified in the lexicon, *Khilōnās*, and card sorting by toy designers; these groups are shared fully or partially with the patrons and shopkeepers.

The centrality of the toy designer in this analysis is essential to the study's aim: to understand the thinking and making of cultural designs by the toy designer as influenced by the perceptions of consumers and shopkeepers. This examination highlights how the toy designer's role is pivotal in the intersection of tradition and contemporary demands and reflects the intricate process of cultural production within the *Bāṇārasī* community.

6.1 Categorization Based on the Meaning of a Khilōnā

The meaning of the *Khilōnā* is well understood by all community members: toy designers, patrons, and shopkeepers of *Bāṇāras*. Here, *meaning* refers to both the connotation of the *Khilōnā* and its intended use. In *Bāṇārasī* culture, the purpose of *Khilōnās* is intrinsically tied to their meaning, which creates a nested relationship between the two. For example, Kṛṣṇa's *Mūrti* (meaning) is intended to be displayed in a *Jhāṅkī* (purpose). This connection between meaning and purpose is reflected in the nomenclature used to describe these groups of *Khilōnās*.

The *Bāṇārasī* community's lexicon reveals three different kinds of nouns to denote the shared meaning of a *Khilōnā*: *Mūrti*, *Khelne kā Khilōnā*, and *Decoration/Sajāvat*. Among these terms, *Khilōnā* alone has a dual meaning. It refers to the entire range of wooden *Khilōnās* produced in *Bāṇāras* and also specifically to those made for children. Let us examine these three groups in detail:

(a) ***Mūrti***: In *Bāṇāras*, *Mūrtis* typically represent divinity, depicting gods and goddesses as couples like *Śiva-Pārvatī, Viṣṇu-Lakṣmī*, and *Rādhā-Kṛṣṇa*, or accompanied by their vehicles, such as "*Śiva* with *Nandī*" or "*Kṛṣṇa* on a peacock." These figurines often illustrate stories from Hindu myths and mythology and are displayed in a *Jhāṅkī* for various religious purposes. While they are treated with respect, they are not consecrated and thus are not worshipped. Some *Mūrtis* also depict common folk and animals to enhance the *Jhāṅkī* display (refer to Fig. 16a).

These figurines are generally small, rarely exceeding a foot in height, although some antique pieces from the past are larger. Each part of the body is meticulously hand carved from solid wood, assembled, and painted. The effort and skill required to carve a *Mūrti* increase its cost compared to other *Khilōnās* in the market. *Bāṇārasī* patrons often purchase them seasonally for religious purposes while modern customers may buy them for nonritualistic functions.

The terms *Mūrti* and *Khilōnā* are often used interchangeably, which highlights the secular status of these *Khilōnās* despite their religious representations. Adults typically handle these figurines, but children may also play with them under adult supervision, with the understanding that they should not be defiled.

An exception to this categorization is the small, hollow, lathe-turned sets of pacifiers, rattles, and spinning tops offered to baby *Kṛṣṇa* during *Janmāṣṭamī*. This recent tradition lacks associated stories or customs and has not yet become part of the community lexicon. Therefore, in Fig. 16a, this exception is grouped with other *Mūrtis* and is shaded in grey.

Overall, the creation and use of *Mūrtis* in *Bāṇāras* reflect a blend of religious tradition and artistic artisanship; these objects serve both devotional and aesthetic purposes within the community.

(b) ***Khelne kā Khilōnā***: *Khilōnās* used solely for the enjoyment of children are simply called *Khilōnās* or *Khelne kā Khilōnā* (refer to Fig. 16b). Unlike *Mūrtis*, children are the primary users of these *Khilōnās*. Although *Khelne kā Khilōnās* are secular, many animal characters, such as cows, elephants, and parrots, appear in both *Khilōnās* and *Mūrtis*. Similarly, representations of common folk, like a *Bāṇārasī* lady, resemble those found in *Mūrtis*. This overlap suggests that, aside from divine deities, *Khilōnās* often borrow their meanings and characters from mythology.

Children engage with *Khelne kā Khilōnās* in various ways. They participate in role-playing games with *Khilōnās* like dolls (*Gudda* or *Gudiyā*) or kitchen sets, the latter of which include items like a rolling pin and board, mortar and pestle, and so on (*Rasoi ghar* with a set of *chaklā belan, kūṭnī*, and *chakkī*). Some *Khilōnās* encourage children to play with the objects themselves, such as spinning tops (*Laṭṭū*), or pull, push, and rotate *Khilōnās* like the *Dāna chiṛiyā* or a *Hilantā* (assorted *Bāṇārasī* pecking *Khilōnās*—birds, elephant, parrot, cow, tiger, horse, peacock, and camel), a *Ghumatoo* (carousel), or a helicopter. Toddlers use *Khilōnās* like a *Jhunjhunā* (rattle) or a *Chusnī* (pacifier). *Khelne kā Khilōnās* are *machine made*, which makes them cheap, affordable, and in demand year-round from both *Bāṇārasī* patrons and modern customers.

The lone exception to the *Bāṇārasī* notion of enculturative play seen in *Mūrtis* or ludic play of *Khelne kā Khilōnās* is the Butler. The Butler has been in production for over a decade and is an assimilation of the counting *Khilōnās* that are widely available in plastic. The Butler stacks ten discs with numbers written on them to form a human figure called a Butler. The educational connotation of the Butler has not been observed in the lexicon of *khel* and Khilōnā discussed in Chap. 2 or in the cultural design card sorting analyzed in this chapter. No separate word has evolved in the *Bāṇārasī* lexicon to accommodate the educational character of this *Khilōnā*. Therefore, the Butler has been added to the other *Khilōnās* in Fig. 16b and is shaded in grey.

(c) ***Decoration kā Khilōnā***: *Khilōnās* consumed by modern customers for display purposes, often without awareness or consideration of their religious or ceremonial function, are referred to by the *Bāṇārasī* community as *Decoration kā Khilōnās* (refer to Fig. 16c). These *Khilōnās* are appreciated for their aesthetic appeal, Indianness, or exoticness rather than for the contextual functions they serve for *Bāṇārasī* patrons.

For example, *flying toys* or nested dolls featuring images of Hindu deities fall into this category. The *flying toys* consists of a deity riding on its vehicle and can be hung from the ceiling with a hook. *Bāṇārasī* patrons do not purchase *flying toys* because their worldview does not permit gods to be hung. According to *Bāṇārasī* beliefs, the *Mūrti* of gods should stand on a stable surface. However, modern customers, both non-*Bāṇārasī* Indians and foreigners, often buy these *Khilōnās* for display in their homes.

Sashi Narayan Agarwal, a shopkeeper, encapsulates this attitude succinctly and notes the divergence between generational *Bāṇārasī* values and modern consumption patterns. For *Bāṇārasī* residents, the spiritual and contextual significance of these *Khilōnās* is paramount, whereas modern customers are primarily attracted to their visual and cultural appeal. According to Sashi Narayan Agarwal, modern clients, particularly foreigners, do not understand the nuances of the *Bāṇārasī* culture.

They take what they understand. If, by chance, they purchase something and cannot afford its rent on the flight, they discard it. They only purchase it for their leisurely pursuit.

Transcription 14: Shashi Narayan Agarwal, shopkeeper at Vishwanath Gali

(a) *Mūrti* (b) *Khelne kā Khilōnā* (c) *Decoration kā Khilōnā*

Fig. 13 Categorization based on meaning. *Image source* Author

Decoration kā Khilōnās, often purchased by modern customers for their visual appeal, can be carved, assembled from flat cut-outs, or turned on a lathe. These *Khilōnās* are characterized by their high skill and finish, which makes them expensive. Some of these *Khilōnās* are unique or produced in limited editions, while others are antiques. The limited editions often include highly skilled carvings, some of which are unique. These *Khilōnās* are closer to being works of art rather than mere objects of play, whether enculturative or ludic.

Older *Mūrtis*, such as those depicting *Viṣṇu*, *Kṛṣṇa*, and *Rādhā*, which have been used in Bāṇāras for religious or ceremonial purposes, have attained antique status due to their age. These antiques are sold to connoisseurs at exorbitant prices. Some toy designers and all shopkeepers have realized that older *Khilōnās* fetch higher prices, leading to a market where even fake *Bāṇārasī Khilōnās* with chipped paint and broken arms are available.

The three boxes of Fig. 13a–c—group *Bāṇārasī Khilōnās* according to their meaning and purpose, along with two deviant members. This categorization, based on the meaning of the *Khilōnās*, is shared among all community members—toy designers, consumers, and sellers—as evident in their lexicon.

Beyond mere meaning, toy designers possess a unique system of categorization that manifests in their specialized vocabulary and mirrors their distinct worldview. This lexicon is employed not only in communicating specific design or manufacturing details among themselves but also in their interactions with shopkeepers. These categories profoundly influence the creation and finishing of *Khilōnās* and ensure that they align with the intended significance or purpose as previously discussed.

6.2 Categorization Based on Representation of a Khilōnā

Representation entails the act of re-presenting an entity that already exists, whether it be a precedent or a source. This re-presentation can vary greatly in its level of exactitude. Certain precedents serve merely as points of departure for new cultural designs and, as such, may bear minimal resemblance to the original. At the other end of this spectrum lie modern client commissions, which demand the utmost precision and fidelity to the original design and leave no room for alteration. In both scenarios, painting holds significant value for the toy designer. This is because, in most *Khilōnās*, meaning is conveyed through the act of painting, and any imperfections in the form can be effectively concealed through this medium (refer to Transcriptions 15 and 16).

Most of the skill is required in painting. In cutting, you can only achieve a shape. It is only with painting that this will look like an elephant.

Transcription 15: Dilip Kumar Prajāpati, toy designer, *Bāṇāras*

You see all this is made out of planks of wood. The design was made on the plank and then cut into a shape. All the planks were then simply joined together. Nothing more. But it is with painting that this is standing out. This is looking like a parrot because of the painting.

Transcription 16: Baichaan Maurya, toy designer, *Bāṇāras*

It is through the act of painting that culturally sensitive features are meticulously distinguished in a *Khilōnā*. Different deities are identified by their painted *tilak*,[3] body color, costumes, and other attributes. Such details are often rendered in accordance with descriptions found in holy texts; for instance, the body color of *Viṣṇu* and *Kṛṣṇa* is blue, while that of *Kali* is black. This cultural worldview rigorously safeguards these features in *Khilōnās* designed for *Bāṇārasī* patrons, in stark contrast to those made for modern customers. As previously discussed, even fake *Mūrtis* are created for modern customers, with tell-tale signs of weathering intentionally painted to appear authentic. Consequently, depending on the modern customer, the cost, and the intended purpose, the toy designer varies the level of representational skill. Within the *Bāṇārasī* lexicon, four distinct kinds of artisanship in representation can be observed.

(a) ***Local kām*** is characterized by the use of predominantly flat colors, a minimal number of lines, and sparse elements of ornamentation. This style employs *chatak* colors—vivid, pure hues—primarily for *Mūrtis* or inexpensive *Khelne kā Khilōnās* bought by *Bāṇārasī* patrons. The wood selected for such *Khilōnās* is typically of lower quality, resulting in forms that are not meticulously finished (refer to Fig. 14a).
(b) ***Sada or Plain kām***, characterized by flat colors with minimal or no ornamentation, is frequently employed for *Khelne kā Khilōnās*, particularly those turned on the lathe, such as a *lattu*, *ghumatto*, or *rasoi saman* as mentioned earlier. Some *Plain Khilōnās* are even left unpainted to showcase the original wooden

---

[3] According to Ajit Kumar Viśvakarma, *tilak* is put on the forehead because the concentration point is the center of the forehead. The *tilak* also describes the caste and the economic status of the people. A single dot denotes a person from a middle-class family such as the mahout on an elephant. Three horizontal lines belong to the Pandit, who hails from a rich class. Chandan is also applied on the forehead to cool down tempers.

(a) Local *kām* (b) Plain *kām* (c) Fancy *kām* (d) *Chaalu kām*

Fig. 14 Categorization based on representation. *Image source* Author

texture, such as the renowned family of saras[4] (refer to Fig. 14b). In contrast to local *kām*, certain *Decoration kā Khilōnās* also remain unpainted to highlight the high level of artisanship and are exorbitantly priced.

(c) ***Fancy kām*** has emerged recently in response to the increasing demand for high-finish, skillfully designed products for modern customers outside *Bāṇāras*. This style elevates the details of well-shaped wooden cultural designs with elaborate painting, often utilizing sombre color schemes and occasionally incorporating *chatak* colors, all executed with fine, intricate brushwork. Within many cultural design communities, including *Bāṇāras*, sombre color schemes are commonly referred to as English colours and are characterized by pastels and lower contrasts. The ornate patterns adorning these *Khilōnās* are meticulously painted following a precise grid.

The term *Fancy*, while often associated with painting, also denotes highly skilled wooden carving or the use of high-quality wood, such as shisham, in manufacturing *Khilōnās*. *Fancy Khilōnās* are more expensive than *Local* or *Plain Khilōnās* and demand greater artistic expertise. There is a growing aspiration among *Bāṇārasī* toy designers to engage in *Fancy kām*, which not only offers higher pay but also provides an opportunity to showcase their artisanship (refer to Fig. 17c).

(d) ***Chaalu kām*** refers to very small, unfinished *Khilōnās* made with minimal care from very low-quality wood, intended for *Bāṇārasī* patrons, and sold at low prices in *melas* and *haats* and by roadside hawkers (refer to Fig. 14d). This type of work is diametrically opposed to the trend of *Fancy kām*. Though not frequently mentioned in the toy designers' vocabulary, *chaalu kām* can still be observed in the market. The reasons for such a decline in artisanship are difficult to pinpoint with certainty. It may be driven by the rising costs of production and by the modern customer's demand for lower prices, especially when compared to plastic alternatives. The cost of wood has increased significantly; wood is now sold by the foot rather than by the pound, as it was 30 years ago. This

[4] The family of Saras was on old *Khilōnā* consisting, as the name suggests, of two large cranes along with two smaller ones. The *Khilōnā* was not painted and was not in production for many years. But recently, it has been revived.

surge in production costs struggles to keep pace with local consumption, which continually bargains for cheaper *Khilōnās*.

In addition to the decreasing *Bāṇārasī* consumption, the recent inclusion of other untrained castes, such as *Mauryas* and *Sharmas*, in the generational toy designer workforce may also impact the quality of the *Khilōnās*. The *Mauryas*, traditionally farmers, have begun to diversify and upgrade their skills into cultural design practice due to decreasing agricultural yields. Consequently, the practice of hereditary skill sets being transferred from older to younger generations is no longer as prevalent.

Thus, the *Khilōnās* organized in Fig. 14 can be categorized into four different boxes if perceived from the point of view of representation rather than meaning. In Fig. 14, *Mūrtis*, *Khelne kā Khilōnās*, and *Decoration kā Khilōnās* can fall in the same group if their features of representation are similar. This indicates a possibility of overlap between the two ways of categorization.

6.3 Categorization Based on Production of a Khilōnā

Three kinds of production skills are frequently referenced in the lexicon of toy designers and shopkeepers: *Khudāi/Gadhwā kām* (carved *Khilōnās*), *Patri/Pattern ke khilone* (assembled flat-*pattern Khilōnās*), and *Kharād ke khilone* (lathe-turned *Khilōnās*). As discussed in Chap. 1, the history of *Bāṇārasī* production skills traces different origins—from the wood-turning toy designers of Rajasthan about 400 years ago to the ivory carvers and clay *Khilōnā* makers of about 100 years ago. These historical roots are reflected in the three distinct production processes still evident in both the lexicon and practice of *Bāṇārasī* toy designers.

(a) ***Gadhwā/Khudāi ke Khilone*** are meticulously hand carved from solid wood. The term *Gadhwā* derives from the word *gadhna*, meaning "to shape," and *Khudāi* from *khodana*, meaning "digging." This intricate and time-consuming work predominantly employs tools such as the *reti*, *aari*, *chausi*, *hathauda*, and *cheni* to carve a *Khilōnā* from a log of wood.[5] Traditionally, this skilled artisanship was reserved for *Mūrtis*, which are costly and purchased with great care by *Bāṇārasī* patrons. Many *Decoration kā Khilōnās* are carved for modern customers who can afford such expensive artisanship (refer to Fig. 15a).

(b) ***Kharād ke khilone*** are primarily hollow *Khilōnās* turned on the lathe. In earlier times, a hand-held lathe was used, but today, the wood is turned on an electric machine with the aid of tools such as the *basula*, *chaursa*, *rukhana*, *barma*, *reti*, *ara*, *baaki*, and *batali*.[6] Many generational designs of *Khilōnās*, such as

[5] *Aari* is used to cut a log of wood into sizes appropriate for the *Khilōnā*. *Chausi*, sometimes accompanied by a hammer, is used to peel of the skin of the wood. *Cheni* is a pointed tool. *Reti* is generally used for harder wood like *gullar*, whereas work on soft wood like *kaima* is done by *Cheni* itself.

[6] Wood is first cut into the required size. A *basula* is used to make a pointed edge on one end which is then hammered on to the machine. As the machine starts, *Banki* is used to sharpen the wood.

(a) Gadhwā kām (b) *Kharād kā kām* (c) Pattern *kā kām*

Fig. 15 Categorization based on production skills. *Image source* Author

spinning tops, rattles, and pacifiers, are turned on the lathe. Nested dolls are also designed using this method (refer to Fig. 15c).

(c) ***Patri/Pattern ke khilone*** are designed using a furma or template, which is cut by machine or by hand on flat wooden plates called Patri. These flat cut-outs are then assembled or joined together to create a *Khilōnā*. This method is commonly employed for making most *Khelne kā Khilōnās* and some *Decoration kā Khilōnās* (refer to Fig. 15b).

Like Figs. 13 and 14, Fig. 15 also distinguishes the production of the same set of *Bāṇārasī Khilōnās* and indicates obvious overlaps of two categories: meaning and representation.

Exploring the common overlaps among these multiple categories can provide deeper insights into the variety that exists within *Bāṇārasī Khilōnās*. All *Khilōnās* in regular production are accepted by the community, which implies that the combination of multiple categories shaping these *Khilōnās* should also align with community-accepted methods of production. Therefore, examining these overlaps could reveal potential combinations for making a *Khilōnā*.

In Fig. 16, *Khilōnās* have been mapped at the intersection points of the three categorization methods, plotted on the *X*, *Y*, and *Z* axes. The *X*-axis represents categorization based on meaning, the *Y*-axis represents categorization based on representation, and the *Z*-axis represents categorization based on production. All of the axes are shown in a one-point perspective. This mapping provides a comprehensive view of the diverse and acceptable ways of creating *Khilōnās* within the *Bāṇārasī* community.

Examining Fig. 16 reveals that some squares are filled while others contain only a single instance of a *Khilōnā* or are entirely empty. According to the reasoning previously discussed, the filled squares represent community-accepted combinations of multiple categories and are populated with various *Khilōnās*. Conversely, the

Using a file, the bark is removed. The untapered side is shaped first into a cylinder. *Barmi* is placed on a firm support and its rounded edge touches the surface of the wood, thus drilling a hole into it. The outer surface is then given shaped and filed to smooth it. The object is colored with a lacquer stick at this point. The completed piece is then removed from the machine.

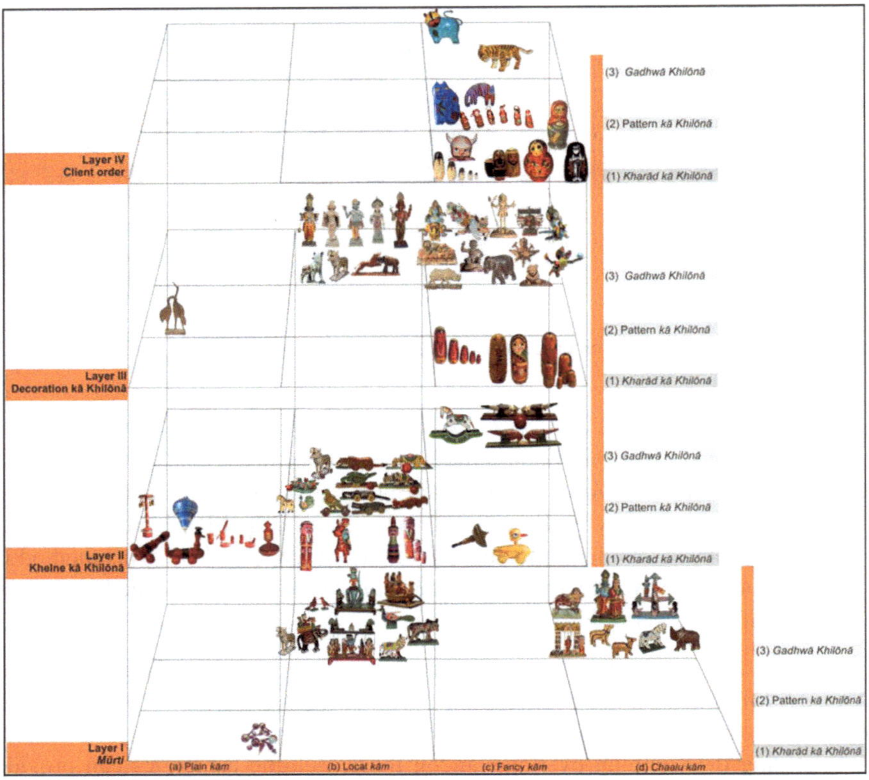

Fig. 16 *Bāṇārasī* community-held categories. *Image source* Author

empty squares indicate culturally unacceptable ways of designing *Khilōnās* within the *Bāṇārasī* community. Therefore, the most favored combinations are those with the highest number of *Khilōnā* instances. However, preferred combinations do not necessarily equate to acceptability within the worldview, as we shall see.

(a) Typically, *Mūrtis* are religious representations of stories used in festivals. They are often carved or made in *Gadhwā* and represented with Local *kām* (refer to Fig. 16, Layer I-b-3).
(b) *Khelne kā Khilōnās* are secular *Khilōnās* with mechanical movement, and they depict animals familiar to the *Bāṇārasī* context. These are generally assembled in patterns and frequently represented with *Local kām* or *Plain* (*sada*) painting (refer to Fig. 16, Layer II-b-2).
(c) *Decoration kā Khilōnās*, which may denote secular or nonsecular themes depending on modern customer preferences, can be produced using *Gadhwā*, *Pattern*, or *Kharād* techniques, but are invariably represented with *Fancy kām* (refer to Fig. 16, Layer III-c-1, 2, 3).

Exceptions to these typical characteristics exist. For instance, *Mūrtis* are generally solid and made in *Gadhwā*, but some *Khilōnās* (see Fig. 16, Layer I-a-1) are turned on a lathe and feature *Sada* or *Plain kām*. The *Bāṇārasī* worldview permits this deviation from the standard of solid, handmade *Mūrtis*. In the past, larger versions of these *Khilōnās*—such as pacifiers and rattles—were purchased for toddlers, but with the advent of cheap plastic equivalents, the wooden versions have lost popularity. Today, miniaturized versions are made as offerings to baby Kṛṣṇa during *Janmāṣṭamī*.

The *Bāṇārasī* worldview is more flexible with *Khelne kā Khilōnās* than with *Mūrtis*. Although typically made with *pattern* assembly and *local kām*, some *Khelne kā Khilōnās* are chamfered into shape on the fret saw, simulating hand carving (refer to Fig. 16, Layer II-c-2). Some are also painted *fancy*, which increases their cost and reduces their appeal to *Bāṇārasī* patrons, thereby decreasing their market presence. Here, toy designers and local shopkeepers take a market risk.

Decoration kā Khilōnās exhibit extensive variation within and between groups, and are open to all categories of meaning and production skills as long as they are Fancy. An interesting deviation includes nested dolls with deity images, such as *Gaṇeśa* and Shiva, which are hollow despite the taboo against making hollow *Mūrtis*. The *Bāṇārasī* worldview accommodates these *Khilōnās* within the flexible category of *Decoration kā Khilōnās*, which are intended for modern customers. This flexibility allows for deviations if the *Khilōnās* are produced for non-*Bāṇārasī* consumption.

Modern client orders produced in Bāṇāras but not adhering to its worldview are made solely for modern customers with little freedom for change or deviation. Examples include cats and Christmas decoration figurines (see Fig. 16, Layer IV).

This categorization provides a fair explanation of community-accepted and less-accepted *Khilōnā* variations. However, is this categorization absolute or static? If static, it would strongly discourage and limit change. But, as seen in Chap. 2, a community's lexicon or vocabulary evolves. Consequently, the meaning of *Khilōnās* has shifted from enculturative to ludic and, in recent years, to display and museumization. As the meaning of *Khilōnās* changes, so do their categories. Based on approximate timelines suggested by toy designers, a timeline of *Bāṇārasī Khilōnā* evolution over the past 100 years has been mapped.

The changing trend from enculturative to ludic to decoration or display observed in the *Bāṇārasī* lexicon is evident in Fig. 17. In the 1900s, the number of *Khilōnās* and categories was small—*Mūrtis* and *Khelne kā Khilōnās*, either carved or lathe turned, with plain or local *kām*. The category of *Pattern kā Khilōnās* emerged between the 1930s and 1960s. Since then, the variety of *Khilōnās* has increased, with new categories like *Decoration kā Khilōnās*, *Fancy kām*, and *Chālū kām* emerging in the mid-1990s to meet the changing preferences of *Bāṇārasī* patrons and the increase in non-*Bāṇārasī* demand. Some *Khilōnās* have shifted categories, like the set of lathe-turned *Khilōnās* offered to Kṛṣṇa, which moved from *Khelne kā Khilōnā* in the 1900s to the *Mūrti* category according to modern customer preferences. *Khilōnās* that went out of circulation, like the Saras family in the 1980s, have re-emerged in the *Decoration kā Khilōnā* category for modern customers. Antiques in the *Decoration kā Khilōnā* category were once part of the *Mūrti* category. Some non-*Bāṇārasī* modern client orders have also been assimilated into *Bāṇārasī Khilōnās*, such as the Russian nested doll and the pecking *Khilōnā* from Eurasia. Despite weakening *Bāṇārasī* demand, generational categories like *Mūrti* and *Khelne kā Khilōnās* persist. However, as *Khilōnās* move away from their enculturative role, *Decoration kā Khilōnās* constitute a larger share of *Bāṇārasī Khilōnās*. Toy designers increasingly take risks designing for exhibition or display rather than generationally enculturative or ludic functions (refer to Fig. 17). With declining *Bāṇārasī* consumption, there's a growing tendency to design for contexts outside the *Bāṇārasī* worldview. The conflict between altered situations and the existing worldview has spawned new categories of production and consumption. The market-induced category of *Decoration kā Khilōnā* emerged in the 1990s to cater to modern customers unaware of *Khilōnās*' enculturative context. This new category also introduced a *Fancy* representation method closer to new consumer tastes.

Several changes can be observed in the *Bāṇārasī* categories and their manifestations as some *Khilōnās* die out, new ones are introduced, non-*Bāṇārasī Khilōnās* are assimilated, and existing *Khilōnās* shift categories to be revived. This evolution occurs despite multiple toy designers working on the same toy and the absence or lack of a singular leader to navigate an increasingly unfamiliar situation. How can we account for the cultural design decisions that have led to such changes?

In the *Bāṇārasī* context, the lexicon shared among toy designers, shopkeepers, and consumers, while exhibiting some variations, generally aligns on the denotation, organization, representation, and description of a *Khilōnā*. This consensus underscores that *Khilōnās*, shaped within a framework governed by a collective worldview, reflects the community-held categorization rather than being the product of an individual creator. The perceptions and inputs of all community members—toy designers, patrons, and shopkeepers—are intricately woven into the design and creation of the *Khilōnā*. Clients, who share this community-held worldview, drive demand based on their specific purposes and tastes as well as their purchasing power. Toy designers not only cater to the modern client's desires but also infuse their creations with the skill and interpretation rooted in *Bāṇārasī* traditions and rituals. Shopkeepers serve as intermediaries, bridging the gap between *Bāṇārasī* patrons and producers. Consequently, the conceptualization and creation of a *Khilōnā* is a communal endeavor—it takes an entire community to think of a *Khilōnā*!

However, the collective community perception of a *Khilōnā*—its meaning and purpose—does not align with the preferences of modern customers from outside Bāṇāras. The modern customer, who does not share the *Bāṇārasī* worldview, often demands *Khilōnās* that conflict with the generational *Bāṇārasī* ethos. This discord leads to disengagement of the toy designer from the cultural designs produced for modern consumption, with both the toy designers and *Bāṇārasī* patrons often refusing to accept these culturally incongruent products. Moreover, toy designers are frequently resistant to making changes to client orders from modern customers for improvement or assimilation, and this resistance further widens the cultural gap.

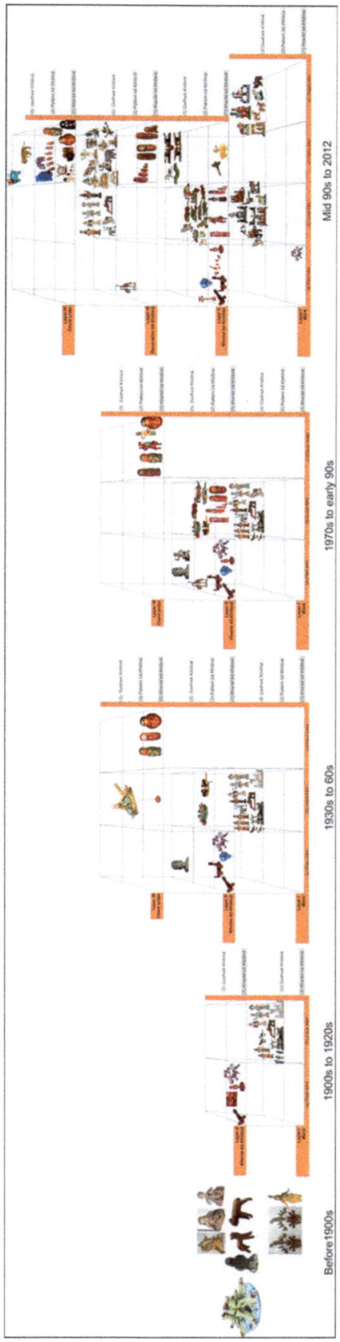

Fig. 17 Evolution of different groups of *Bāṇārasī Khilōnā*. *Image source* Author

References

Alexander C (1964) Notes on the synthesis of form. Harvard University Press
Athavankar UA (1989) Categorization… natural language and design. Des Issues 5(2):100–111
Coomaraswamy AK (1909) The Indian craftsman, vol 1. Probsthain & Company
Cross N (2006) Designerly ways of knowing. Springer, London
Lakoff G (2008) Women, fire, and dangerous things: what categories reveal about the mind. University of Chicago Press
Jones, JC (1970) Design methods: seeds of human futures. Wiley, New York
Krippendorff K (2006) The semantic turn: a new foundation for design. CRC Press, London
Miller D (1985) Artefacts as categories: a study of ceramic variability in Central India. Cambridge University Press
Rosch E, Lloyd B (1978) Cognition and categorization. Lawrence Erlbaum Associates, Hillsdale, NJ
Serena N, Warms RL (2007) Cultural anthropology. Thomson Wadsworth Corporation

Chapter 4
View of a *Khilōnā*: Rules of Making and Using Cultural Design

1 The Empty Boxes in the Category Cube

Among the several endearing folk toy designs that have engaged children and adults alike is the *(Dānevālī chiṛiyā)*. The design of the *(Dānevālī chiṛiyā)* diffused far and wide in countries spread across all the continents, so its origin is difficult to trace. However, the oldest documented instance of an Indian *Dānevālī chiṛiyā* probably comes from a small collection of *Khilōnās* owned by Lovat Fraser, an artist who passed away in 1921.

In his seminal essay, "Pecking Chickens: A Folk Toy as a Source for the Study of Worldview," Alan Dundes (1989, 83–91) examined pecking toys from India, Italy, Poland, Spain, Russia, and the United States and deduced patterns of food consumption and spatial arrangements within the cultures the toys originated from. Dundes observed that Polish chickens, pecking from a single communal bowl, which was nearly empty, had very little space between their individual heads. This depiction, he argued, was a reflection of the communist government in Poland from the 1940s to the 1980s. Conversely, the use of real kernels of corn in these toys indicated an abundant food supply, suggesting that only a country with surplus food would use it to decorate a toy. Upon examining the Italian toy that included an egg, Dundes postulated that the toy resembled the mother–child relationship, which is a prominent feature in Italian culture.

However, the interpretation of the *Bāṇārasī Dānevālī chiṛiyā* intrigued me the most. Dundes hinted that the eight chickens pecking on a figure, without any actual food but rather a pretty scenery, indicated the population explosion in the subcontinent and the corresponding difficulties in ensuring food security. He speculated that in

---

This chapter draws upon selected material from the author's earlier publication, "*Dincs as Worldviews: Things that Communicate a Mind*," in Toys and Communication, pp. 137–152, London: Palgrave Macmillan UK, 2017. The current version has been revised for clarity and expanded in scope to include the conventions, canons, and beliefs surrounding a wider range of *Bāṇārasī Khilōnās* beyond the pecking toy.

K. Patil, *Toying with the Mind*, Design Science and Innovation,
https://doi.org/10.1007/978-981-96-9227-9_4

Indian culture, art or cosmology might serve as a worthy substitute for food, given the absence of actual food items in the toy.

Despite Dundes's frequent cautions against making broad speculations from limited examples, he attempted to explicate the implicit social constructs embedded within these toys by their makers. Food and space are indeed critical matters of concern and are enthusiastically negotiated and debated within any worldview. Hence, it is not surprising that these concerns are reflected in the microcosm of a toy. However, as an ethnographer deeply engaged in the study of *Bāṇārasī* designs, I find Dundes's observations inadequate to explain the rich complexity of the *Bāṇārasī* mindset and its expressions, as exemplified by the *Bāṇārasī Dānevālī chiṛiyā.* My reluctance to accept Dundes's explanation stems from the complexity of the category cube discussed in Chap. 3.

While vocabulary reflects categorization within the culture, the objects themselves are subject to multiple ways of categorization by the *Bāṇārasī* community. Only certain combinations of the multiple categories of meaning, representation, and production, as shaped by the worldview, can produce *Khilōnās* that are acceptable to the *Bāṇārasī* context. This implies that the relationship between these categories is held constant at a given point in time by the worldview, which determines and limits the acceptability of design variations by the community. The permissible combinations of categories determine the variety or lack thereof in *Bāṇārasī Khilōnās*, making it crucial to understand the rules by which some combinations are accepted while others are rejected.

The combination of multiple community-held categories has produced an immense variety of *Bāṇārasī Khilōnās*. For instance, all the tigers in Fig. 1 are *Bāṇārasī Khilōnās*, which are produced within the boundaries of different community-held categories. Yet there is a notable variety of *Khilōnās* available for the same meaning but in different categories. The tiger, considered one of the favorite animals of *Thākurjī* and a vehicle of Goddess *Durgā*, is conventionally used in a *Jhāṅkī* for display or along with the Goddess herself. The *Khilōnā* in Fig. 1, layer I-b-3, is an old hand-carved tiger painted in a local style for a *Jhāṅkī*. Discarded because it is old, it can be sold as an antique artefact today, which places it in both the *Moorti* and *Decoration kā Khilōnā* categories. Another example, Fig. 1, layer III-b-2, is a tiger made from *pattern* assembly with local painting and designated for children's play, which categorises it as a *Khelne kā Khilōnā.*

However, not all combinations of categories are accepted by the worldview. Figure 1, layer IV-c-3, features a tiger, but toy designers have not retained it as a *Bāṇārasī Khilōnā* despite having the skill to produce it. This *Khilōnā* was made for an export order and was not reproduced again by the toy designers, nor did the local shopkeepers place orders for it again. The *Bāṇārasī* worldview does not accept representations that are caricatures or cartoons; hence, this exaggerated tiger is considered unacceptable.

Despite the large variety of tiger *Khilōnās*, Fig. 1 still has empty squares. This implies that certain combinations of categories are unacceptable to the community worldview, which makes it impossible to conceive of any artefact within those parameters. For instance, Fig. 1, Layer I-a-1 (*Mūrti*) is empty because the worldview does

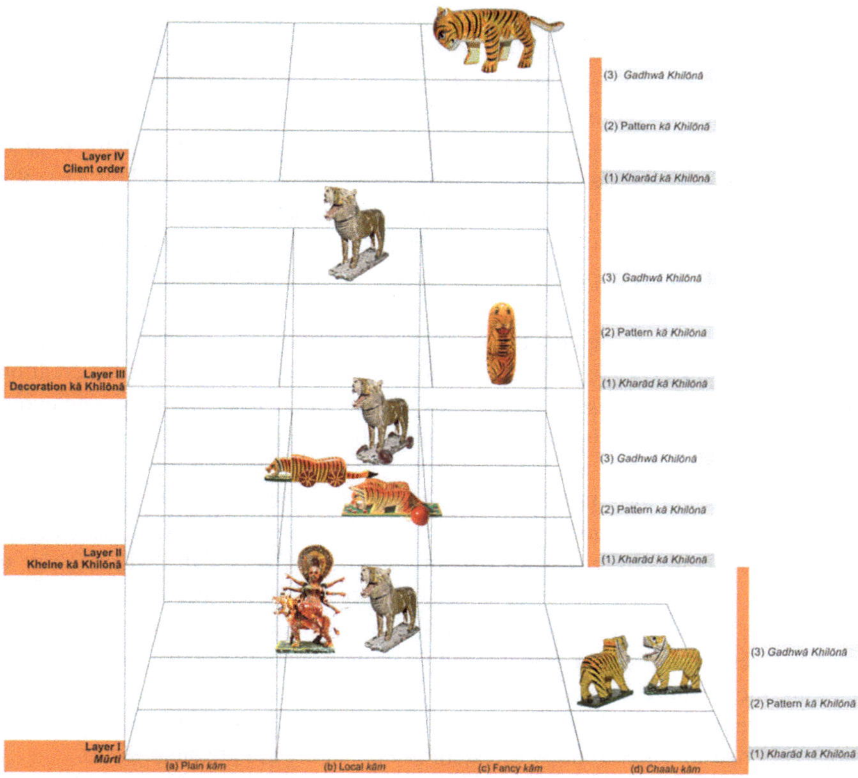

Fig. 1 Organization of various tigers in community-held categories. *Image source* Author

not permit a hollow machine-made *Plain Khilōnā* to be placed in a *Jhāṅkī*. Understanding the rules by which some squares are populated and others are empty is crucial for grasping the acceptability of *Khilōnās* within the *Bāṇārasī* community.

But what are these rules that govern design decisions of cultural designers, and how does one elicit them?

Moalosi et al. (2010) observed that current design methodologies, with their rigid standards, rules, and guidelines, often fail to address cultural contexts adequately. In the absence of a solid theoretical and cultural framework, designers frequently mimic Western concepts and overlook the local context. This disconnect is also evident in understanding the thought processes of India's culturally sensitive design communities. While the manufacturing details of these communities are well documented, the crucial unwritten standards, rules, and guidelines regarding the design and use of these artefacts remain elusive.

In examining the intrinsic, evolutionary nature of community practices, it becomes evident that the need for consensus, local legitimacy, and the tacit endorsement of worldviews bind habitual actions to community-practiced conventions, canons, and beliefs. Unlike autonomous, consciously established standards, rules, and guidelines, these cultural constructs are integrally woven into the social fabric, much like the creation of a *Khilōnā*. Unlike modern *Khilōnās*, which are often confined to specific users (such as children) and specialized contexts (like a playhouse or playground), a *Khilōnā* transcends these boundaries. Similarly, conventions, canons, and beliefs extend beyond the mere creation of a *Khilōnā*; they are interwoven with the broader community practices of ceremonies, rituals, festivals, and daily life. This interplay not only situates the *Khilōnā* within a particular cultural context but also reflects the community's worldview. Thus, the *Khilōnā* becomes an enculturative expression of *Bāṇārasī* culture as opposed to a universally designed object devoid of cultural significance.

Conventions, canons, and beliefs shape the community's perspective on artefacts, particularly *Khilōnās*. They guide the toy designer's decisions regarding the creation, use, and acceptance of these artefacts, thus influencing both their variety and evolution. To fully understand these decisions, it is crucial to define conventions, canons, and beliefs.

The literature on the role of conventions, canons, and beliefs in cultural design communities, and even in modern design, is sparse. However, some studies do touch upon these concepts. In the context of authored design, especially with the rise of user-centric approaches, conventions are occasionally referenced. Norman (1999, 38–43) describes conventions as cultural constraints shared by a group. They evolve over time, restricting certain activities while promoting others. Conventions are not arbitrary; they require a community of practice to develop and are slow to adopt but hard to discard once established. Norman further warns that violating conventions can greatly disrupt a community's system of navigating the physical world.

Similarly, the concept of canons has been extensively explored in art and anthropology. A prime example is the canons of proportion evident in Leonardo da Vinci's Vitruvian Man, which was based on the work of the architect Vitruvius. These canons establish the ideal proportions of different body parts, with the human body standardized to eight heads in height, and connect this ideal to principles of architecture. Likewise, Tschichol's "Golden Canon of Page Construction" and the Van de Graaf Canon provide rules for dividing a page into aesthetically pleasing book layouts. In Coomarswamy's collection of essays, edited by Lipsey (1977: 80, 113), Coomarswamy elaborates on the Indian perspective of *Pramāṇa* and states that this perspective parallels the Western notion of a canon. Toy designers are expected to internalize the knowledge of *Pramāṇa*, which encompasses ratio or proportion and is subdivided into canons such as *naya* (means), *vidhi* (rite and ritual), and *mana* (measurement), as recorded in technical texts like the *Śilpaśāstra*. In this sense, *Pramāṇa* represents the form or embodiment of knowledge itself.

Beyond the visible manifestations of conventions and canons in artefacts, there exists an invisible layer of beliefs. Katz (1960, 168) discusses these beliefs as descriptions and perceptions of an object, its characteristics, and its relationships with other objects. Leung et al. (2002) further describe beliefs as basic premises that guide behavior across different situations. Lin (2007) includes beliefs in his cultural model for design; he identifies it as beneath the tangible visceral and behavioral design features and as embodying values, emotions, and beliefs. He references the spiritual or ideal level of Lee (2004) and the intangible level of Leong and Clark (2003) within this framework. Similarly, anthropologist Redfield (1941, 133) refers to "conventional understandings manifested in act and artefact that persist through tradition," where understandings encompass beliefs, meanings, values, and other mental constructs related to the artefact. In un-self-conscious cultures, Alexander (1964, 48) considers these mental constructs or the invisible layer to be an ingrained fixity of patterns of myth, tradition, and taboo that resist deliberate change. Given the varied nomenclature in design literature for this mental or emotional level, this study will adopt the general term *belief* to encapsulate these concepts.

A radical shift in the discussion on conventions, canons, and beliefs is introduced by Ruth Millikan (2005), who emphasizes the role of precedent. Drawing inspiration from biology, Millikan defines *convention* as a pattern of behavior reproduced within a population largely due to the weight of precedent. Unlike consciously designed artefacts, precedent significantly influences culturally sensitive designs, which are widely believed to be built on traditional foundations, such as the shawls of Slovakian peasants (Alexander 1964, 53). The Slovakian peasants relied so heavily on traditional designs that, once the precedent link was disrupted by modern technological pressures, they were unable to achieve the same quality in their shawls despite generations of practice.

In light of Millikan's perspective, it is essential to summarize our definitions of conventions, canons, and beliefs, and to consider the crucial role of precedents in cultural design communities. Conventions are a set of dictums that collectively form an appropriate "know-what" and habitually guide toy designers in making, using, and evaluating an artefact based on the weight of precedent. They outline what is and isn't permitted within cultural design communities and serve as simple assertions or denials regarding the attributes of artefacts. For example, all *Mūrtis* are carved in solid wood; hence the tiger in Fig. 1, level I-b-3, is identified as a *Gadhwā Khilōnā*. Over generations, the origins of many conventions, like those of beliefs are often forgotten within oral community practices. Consequently, Indian toy designers practice conventions without necessarily understanding the rationale behind them.

Canons are community-practiced norms of "know-how" that dictate the manner of representing typical or frequently repeated visible motifs, shapes, or patterns shared among all artefacts within the community due to the weight of precedent. The frequent recurrence of an element, not only in an individual's artisanship but also among all community members, helps identify it as a canon. Canons are domain-specific directives about an artefact's physical and tangible features, such as visual appearance, size, shape, dimensions, manufacturing, and mechanism. Some canons are explicitly mentioned in older texts, like the *Pramāṇa* (proportions) in temple architecture, while others are passed down orally or by emulating a master. For instance, the tiger in Fig. 1, layer I-b-3, features colours that represent a real tiger because arbitrary colours are not used in *Bāṇārasī Khilōnās*. Additionally, *Bāṇārasī Khilōnās*, especially those in the *Mūrti* category, always have a base. This is evident as all other tigers in Fig. 1 also have similar bodily colours and a base, except for the nested doll and client orders in layer IV, which are not governed by the *Bāṇārasī* worldview.

Beliefs are worldview-ascribed values and conceptions that govern relationships between people and, in this context, between people and artefacts. These beliefs are expressed through social and religious values, taboos, or superstitions, largely influenced by precedent, and they shape the consumption and production of *Khilōnās*. Beliefs are resistant to change or replacement, and an artefact that contradicts the community's beliefs can become unacceptable. In India, some beliefs are found in old scriptures and texts, but the origins of most are unknown to the common people. Nevertheless, their oral transfer and observations and imitation by children in their households and communities ensure their continuity. For instance, the tiger in Fig. 1, layer I-b-3, is considered sacred because the *Bāṇārasī* community believes it is a favored animal of *Thākurjī* and the vehicle (*Vāhana*) of Goddess *Durgā*. Therefore, the tiger *Mūrti* is treated with care and respect by both older and newer generations of producers and consumers.

Is there a method to elicit the tacit knowledge of conventions, canons, and beliefs embodied in the practice of toy designers? Methods like protocol analysis and retrospective elicitation of thinking have been used in studies of cultural design thinking, but designers are often verbally articulate and aware of explicit rules, standards, and guidelines, unlike toy designers. In the absence of such cognitive methods for eliciting the tacit knowledge of toy designers, their material culture became the primary source for decoding cultural design thinking in this study.

In the forthcoming sections, we will discuss the method of elicitation used and the conventions, canons, and beliefs derived from this approach. By closely examining the artefacts themselves, we can infer the underlying principles and cultural norms that guide their creation, thus unveiling the often unspoken knowledge that toy designers embed in their work. This material culture approach allows us to reconstruct the thought processes and cultural frameworks that shape design practices and provides a deeper understanding of the conventions, canons, and beliefs integral to their practice.

2 Eliciting Conventions, Canons, and Beliefs

Artefacts are embodiments of human thought, and they reflect the use, the user, and the artefacts' utility within a culture. As products of the human mind, they manifest the thinking of their creators. Understanding an artefact, therefore, equates to understanding its designer. Cross (1982) emphasizes that the knowledge embedded in an artefact represents a significant branch of designer ways of knowing.

Like modern designers, *Bāṇārasī* toy designers think through their artefacts, often bypassing the mediation of paper drawings or detailed planning. This suggests that cultural design thinking, much like modern design thinking, is embedded within cultural designs. In this context, the *Khilōnā* is treated as evidence that not only witnesses but also manifests the design thinking and decisions of a community.

This study examines this thinking indirectly through the decisions embodied in the physical form of the *Khilōnā*. Verbal elicitations from toy designers, as discussed in Chaps. 2 and 3, further corroborate the thinking embodied in the artefact.

For this study, nearly 150 *Khilōnās* were collected through direct purchases from shops, antique warehouses, and toy designers themselves. In addition, objects archived at the Museum of Bāṇāras Hindu University and the Craft Museum in Delhi, India were photographed. These *Khilōnās* span different periods from the 1900s to 2011 and have been approximately sequenced according to the oral history narrated by toy designers. Contemporary *Khilōnās* were easier to place on the timeline because they were readily available on shop shelves.

This collection and analysis of *Khilōnās* serves to decode the tacit knowledge of conventions, canons, and beliefs, and provides a window into the cultural and cognitive processes of the toy designers who created them. By examining these artefacts, we can reconstruct the thought processes and cultural frameworks that underpin design practices and reveal the deep-seated norms and principles that guide their practice.

From this pool, forty *Khilōnās* were selected for analysis and represented different community-held categories: ten each from *Mūrti, Khelne kā Khilōnā, Decoration kā Khilōnā*, and client orders. The selection was based on several criteria to ensure a comprehensive understanding of the *Bāṇārasī Khilōnās*:

- *Typical and atypical instances*: To cover a wide variation, both typical and atypical instances of each category were included. This approach ensures that deviations are considered and provides an informed view of the diversity within *Bāṇārasī Khilōnās*.
- *Visible changes over time*: Instances that exhibited visible changes over a period were chosen to capture the rules that remain valid despite unavoidable transformations. This helps in understanding the natural evolution of even the most tradition-bound designs.
- *Known or identifiable sources of change*: Designs with known or identifiable sources of change were selected to decode these changes and measure their extent through comparison.

By analyzing the designs through these lenses, this study uncovers the design decisions and thinking embedded within the physical manifestations of the *Khilōnās* and offers insights into both the continuity and the evolution within the *Bāṇārasī* community's design traditions.

A systematic two-step approach with three substeps was followed to maintain rigor while decoding the canons, conventions, and beliefs manifested in a design. A feature analysis was conducted to deduce a pattern through a rigorous, iterative process of comparing similarities and differences between features.

(1) Feature analysis: Features of every *Khilōnā* were segregated, coded, and analysed separately. For the purpose of coding, every *Khilōnā* is referred to as an artefact and is *labelled* Artefact of Analysis 1, Artefact of Analysis 2, or AoA3, AoA4, … AoA40. Feature, here, is a distinctive element of a *Khilōnā*

that allows itself to be modified in varying degrees and sometimes relinquishes itself completely in successive iterations. Its external representation facilitates its coding through any means—visual, written, or verbal. Such features related to the artefact (visual, shape, and size), its production (manufacturing, mechanism, and material), and consumption (purpose and connotation) were studied meticulously to arrive at patterns of frequently occurring features (refer to Fig. 2).

(2) Feature analysis was used to deduce a pattern through a rigorous to-and-fro iterative process of comparison of similarities and differences between features. A systematic two-step approach with three substeps was followed to maintain rigor and decode the views gathered and manifested in the artefact.

Step 1: The analysis of a single feature of the instance was written in the form of a statement. The simple statement included a subject and a predicate, where the subject asserted a particular attribute in its predicate. For instance, the *Dānevālī chiṛiyā* (subject) has *round eyes* (predicate). This statement 1 was derived from R1 of Feature Analysis of AoA8 (refer to Fig. 2).

Step 2: Every new feature in the analysis of this table or the forthcoming ones was corroborated with statement 1 to determine the following:

(a) If it asserts statement 1.
(b) If it denies statement 1.
(c) If it is a new statement; that is, both the subject and the predicate are different from Statement 1.

Step 2a: If the new subject asserted the predicate of the statement, it was added to it. Therefore, parrots and all other unidentified bird deities were subsequently added to Statement 1. Thereon, the revised Statement reads as *The Dānevālī chiṛiyā, parrots, and all other unidentified birds* (subject) *have round eyes* (predicate).

Step 2b: The subject was rejected if many similar subjects negated the attribute. However, rare aberrations were considered an exception. For instance, the *peacock* (subject) has *Kamalanayanī* (lotus-like) *eyes* (predicate). This became an exception to Statement 1.

Step 2c: If the new feature had a different subject and attribute from Statement 1, then the process was restarted from Step 1 with a new statement.

Step 3: The statements so derived were coded together as conventions, canons, and beliefs.

Every statement was corroborated with all forty selected toys unless it was strongly refuted by several other instances. Each of the 40 selected toys became Artefacts of Comparison for AoC for the Artefact of Analysis, that is AoA.

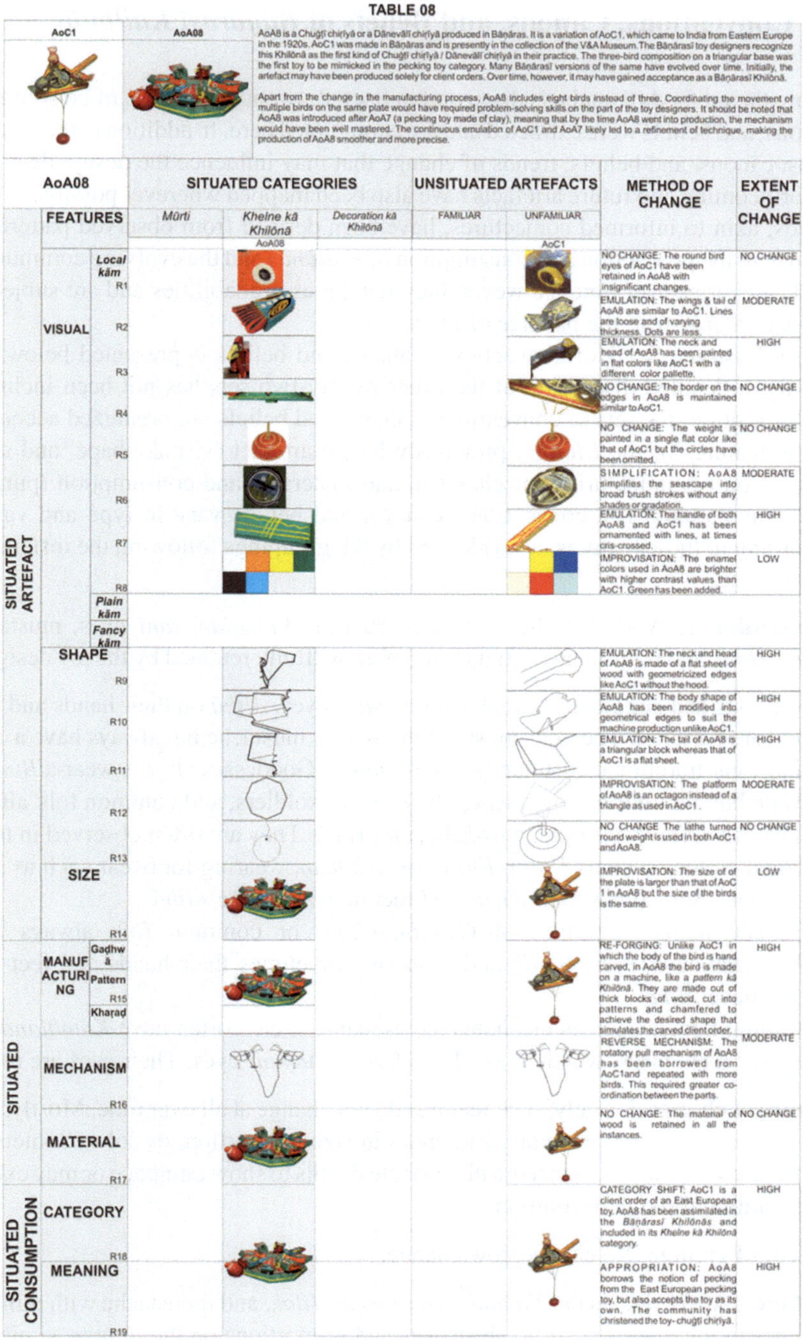

Fig. 2 Feature Analysis of the *Bāṇārasī* Pecking Toy or *Dānevālī chiriyā*. Image source: Author

3 Conventions, Canons, and Beliefs of *Bāṇārasī Khilōnās*

Using the method of analysis discussed in Sect. 2, twenty-three sets of conventions, canons, and beliefs were extracted and are enumerated here. In addition to the conventions, canons, and beliefs, trends of change that may influence the design decisions of forthcoming and future artefacts have also been mapped wherever possible. These trends, akin to informed conjectures, have been derived from observed patterns of change in the production and consumption of *Khilōnās* and the evolving community-held categories over time. However, they remain as probabilities and are subject to validation only with the passage of time.

To illustrate, Set 1 of conventions, canons, and beliefs is presented below. For brevity, a detailed discussion of the other twenty-two sets has not been included. In this section, the sets of conventions, canons, and beliefs are organized according to the features of the *Khilōnā* previously listed: artefact (visual, shape, and size), production (manufacturing, mechanism, and material), and consumption (purpose and connotation). The conventions, canons, and beliefs vary in type and variety according to these features, as evidenced by the groupings following the instance of the first set.

Convention 1: Worldview-held features, such as *Kamalanayanī* eyes, mustache, *Tilak*, *Ālthā*, and traditional costumes are often willfully retained by the toy designer.

> All gods and goddesses have *Kamalanayanī* eyes, *Ālthā* on their hands and feet, and are barefoot. The gods mostly don't have a mustache but always have a *Tilak* and wear traditional costumes such as *Dhotis*. Goddesses always wear a *Bindi*.
> Male human figurines of *Gwālā*, *Dvārapālas*, soldiers, and common folk always have *Kamalanayanī* eyes, mustache, and *Tilak*. They are often observed in traditional clothes such as *Dhoti*, *Kurta*, or *Āchakan*, wearing footwear such as *Jūṭīs* or boots. Sometimes their hands and feet may also have *Ālthā*.
> Female human figurines of *Gwalin*, *Gopī*, or common folk always have *Kamalanayanī* eyes, *Bindi*, and a *Saree*. Sometimes their hands and feet may also have *Ālthā*.
> Animal figures such as elephants, cows, camels, etc., often have *Kamalanayanī* eyes, except birds, which do not have *Kamalanayanī* eyes. Their eyes are round.

Nature of change: Largely, such features do not change at all over time. Mostly, they are emulated with incremental differences in size, proportion, or color. Sometimes such features are also enhanced with intricate details to show opulence or may exhibit deterioration through devolution.

Extent of change: No change, low change.

Instance: AoA1 retains the *Kamalanayanī* eyes, *Tilak*, and moustache with minimal changes in the thickness of brush strokes and proportions on the figures, similar to AoC1 and AoC2. The *Kamalanayanī* eye is also retained on the royal elephant. The traditional costumes of the king and *Mahout* are decorated with three-dot patterns, flowers, circles, and borders to make the artefact "*fancy*" (refer to Fig. 3).

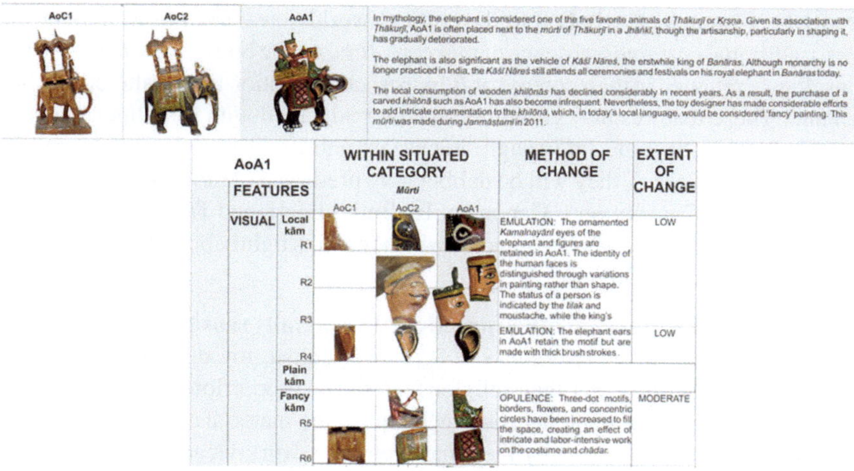

Fig. 3 Illustration of Convention 1. *Image source* Author

Exceptions to the rule: Men in the old *Band-bājā Khilōnā* and *Sukhī Parivār* wear Western clothes with a tie. Birds have round eyes.

Canon 1a: *Kamalanayanī* eyes are generally a big white boat-shaped form with a black full or half circle representing a pupil in between, sometimes spotted with a highlight. The eyebrow is a curved single stroke of black at a distance from the eye.

Canon 1b: Faces do not wear any makeup such as rouge or lipstick.

Canon 1c: The eyes of an elephant and a peacock have an extra swirling line common to body decoration.

Canon 1d: All traditional costumes, such as *Dhoti*, *Kurta*, *Āchakan*, or a *Saree*, have borders. Costumes are mostly made in red, yellow, and orange.

Canon 1e: Men wear *Jūṭīs* but deities are barefoot.

Canon 1f: Women are depicted with a *Pallu* on their heads.

Belief 1a: The moustache signifies that the father of the person is still alive.

Belief 1b: The *Tilak* signifies the social and religious status of a man.

Belief 1c: A *Bindi* signifies the marital status of the woman. Unmarried women are not represented in the *Khilōnās*.

Belief 1d: Hands and feet of all figurines are painted red with *Ālthā*, which is considered auspicious.

Belief 1e: The *Kamalanayanī* eyes are mentioned in the *śāstras* as beautiful eyes, often associated with the deities.

Belief 1f: Women wear a *Pallu* on their heads.

Trend 1a: Features closely tied to the cultural worldview—such as *Kamalanayanī* eyes, traditional costumes, absence of makeup, *Ālthā* on the body, and others—will be consciously retained in their context of *Bāṇārasī* consumption unless altered by client demand. These features may either become more detailed with finer lines in expensive artefacts or degrade into rough, haphazard motifs in lower-income contexts. However, in both cases, they will be deliberately preserved. Artefacts adhering to the aforementioned Convention 1, Canon 1a–1d, Belief 1a–1f, and Trend 1a are predominantly found within the *Mūrti* category, which remains tightly bound to features that reflect the cultural worldview.

Worldview-held features in other practices: In culturally sensitive design communities such as that of the *Aṣṭādhatū Mūrtis* in Ayodhya, toy designers steadfastly refuse to create figures other than religious ones. The proportions of the figures, the colors of the dress, the manufacturing techniques, the materials, the meanings, and the categories are all dictated by the Hindu religious worldview. Men are depicted with square shoulders, broad chests, slim waists, and slightly protruding stomachs, while women, as sustainers of life, are portrayed with rounded breasts and large hips. The rough figure cast undergoes polishing and finishing. The clothes are painted first: *Rāma* wears a yellow *Dhoti*, whereas *Sītā* is adorned in a red *Saree*. Hands and feet are painted with *Ālthā*, and the forehead features a *Tilak* or a *Bindi*, depending on the gender. The baby *Rāma*, or Lalla, holds a *Laddu* in his right hand. The *Kamalanayanī* eyes of the sculptures are fixed with conch shells, which maintain their pearliness even as the paint dulls over the years. The durability of the sculpture is a crucial aspect of its production (refer to Fig. 4).

Even when worldview-held features are produced through non-*Bāṇārasī* or modern production, these sensitive features must be respected and emulated for the artefact to be accepted by *Bāṇārasī* consumers. For example, Chinese manufacturers produce *Mūrtis* of Gaṇeśa for the Indian market. These *Mūrtis* replicate the essential worldview features, such as the *Tilak*, *Kamalanayanī* eyes, and specific colors, despite the manufacturers' lack of understanding of their significance. If these features are not accurately mimicked, the consumers reject the artefact.

All twenty-three sets of conventions, canons, and beliefs are enumerated below under three groups: artefact, production, and consumption. To explain the design decisions behind the variety of tigers in Fig. 1, every set is also followed by an instance from Fig. 1 that either confirms or negates the set wherever found relevant.

3.1 Conventions, Canons, and Beliefs: Artefact

A large number of conventions, canons, and beliefs that represent the features of a *Khilōnā* are grouped under the artefact set. Within this set, features related to visual representation of the artefact constitute the larger share of conventions, canons, and beliefs compared to those of shape and size. Interestingly, while conventions and beliefs appear less flexible in their interpretations, the canons manifest them through

Fig. 4 *Aṣṭādhatū Lalla* and *Lalli* from Ayodhya. *Image source* Author

a variety of representations. This results in a range of visual expressions despite the narrow boundaries set by the conventions and beliefs, which may explain the wide variety in *Bāṇārasī Khilōnās* despite the hold of rigid traditions and taboos. This also supports the claim of toy designers that a *Khilōnā* attains its *Bāṇārasī* identity primarily through its painting or visual representation.

3.1.1 Artefact: Visual Features

Convention 1: Worldview-held features such as *Kamalanayanī* eyes (with lotus-shaped eyes), moustache, *Tilak* (a sectarian mark worn on the forehead chiefly by Hindus), *Ālthā* (a red dye applied to the hands and feet of women in India), and traditional costumes will be largely retained willfully by the toy designer.

- All gods and goddesses will have *Kamalanayanī* eyes, *Ālthā* on their hands and feet, and are barefoot. The gods mostly don't have a moustache but will always have a *Tilak* and wear traditional costumes such as *Dhotis* (a loincloth worn by men in some parts of India). Goddesses will always wear a *Bindi* (a red dot applied on the forehead between the eyebrows to symbolise marital status).
- Male human figurines of *Gwālā* (a person who attends to cattle), *Dvārapālas* (an armed guardian who secures the gates, often of a temple or castle), soldiers, and common folk will have *Kamalanayanī* eyes, a moustache, and a *Tilak*. They are often observed in traditional clothes such as *Dhoti*, *Kurta*, or *Āchakan*, wearing

footwear such as *Jūṭīs* or boots. Sometimes their hands and feet may also have *Ālthā*.

- Female human figurines of *Gwalin* (a female, often the wife of a *Gwālā*, who attends to cattle), *Gopī* (female cowherds, who are enamoured with Lord *Kṛṣṇa* and dance with him and perform a *Rāsleelā*), and common folk will always have *Kamalanayanī* eyes, *Bindi*, and a *Saree*. Sometimes their hands and feet may also have *Ālthā*.
- Animal figures such as elephants, tigers, cows, camels, and so on, will often have *Kamalanayanī* eyes.
- Birds will not have *Kamalanayanī* eyes but will have round eyes.

Canon 1a: *Kamalanayanī* eyes are generally a big white boat-shaped form with a black full or half circle representing a pupil in between, sometimes spotted with a highlight. The eyebrow is a curved single stroke of black at a distance from the eye.

Canon 1b: All traditional costumes, such as *Dhoti*, *Kurta*, *Āchakan*, or a *Saree*, have borders. Costumes are mostly made in red, yellow, and orange.

Canon 1c: Faces do not wear any makeup, such as rouge or lipstick.

Canon 1d: Men wear *Jūṭīs*, but deities are barefoot.

Canon 1e: Women are depicted with a *Pallu* on their head.

Canon 1f: The eyes of an elephant and a peacock have an extra swirling line common to body decoration.

Belief 1a: The moustache signifies that the father of the person is still alive.

Belief 1b: The *Tilak* signifies the social and religious status of a man.

Belief 1c: The *Bindi* signifies the marital status of the woman. Unmarried women are not represented in the *Khilōnās*.

Belief 1d: The hands and feet of all figurines are painted red with *Ālthā*, which is considered auspicious.

Belief 1e: The *Kamalanayanī* eyes are mentioned in the *śāstras* (a treatise or text, often of the Vedic period on a specific field of knowledge) as beautiful eyes, often associated with the deities.

Convention 2: A contrasting dash of color is often added to a bird's wings as an aesthetic device to break the monotony of the texture of the wings.

- All *parrots* will have a red blotch on their wings.
- *Peacocks* often have a blotch on their wings.
- *Dānevālī chiṛiyā* generally do not have a blotch on their wings.
- Unidentified birds may not have a blotch on their wings.

Canon 2a: Often, a complementary or high-contrast color is used to make a swift, small brush stroke on the wing of a bird.

Canon 2b: Cows and horses do not have a blotch but a repetition of circles.

Belief 2a: The *sugga* or the parrot is revered as one of the favorite animals of *Thākurjī* (another name for Lord *Kṛṣṇa*, who is considered the head of *Vṛndāvan)*. The red patch is traditionally added to the green wings of the *sugga*. The origin of this practice is not known. It may have been adopted in *Khilōnās* from the practice of putting vermilion on images. It may also have its origin in protection against the evil eye.

Belief 2b: Feeding birds on the *Ghāṭs* (a set of steps leading down to a river, lake, or temple pond) and owning a parrot at home are considered pious actions in Bāṇāras.

Convention 3: *Chatak* colors will be willfully used by toy designers to visually represent a *Bāṇārasī Khilōnā*. Such colors are bright and are often placed in combinations that enhance their contrast values.

Traditionally, golden yellow, rusty red, emerald green, creamy white, dull black, and earthy brown were mostly used in the *Khilōnās*. Golden and silver colors have often not been a part of the *Bāṇārasī* color palette. In the past, yellow often replaced golden color in jewelry.

Canon 3a: The costumes of deities are largely rendered in yellow and red.

Canon 3b: Colors are a device to represent the real world unless the meaning is framed from mythology. Therefore, a parrot is green, a cow is white, an elephant is black, a camel is ochre yellow, and a tiger is yellow.

Canon 3c: Ludic *Khilōnās* like *Laṭṭus* or spinning tops that do not represent any particular meaning can be painted in any bright, warm color.

Canon 3d: In the absence of golden and silver colors, jewelry is painted in yellow and white, respectively.

Canon 3e: A flat base color is applied on most *Khilōnās*. Motifs are painted after the base layer has dried. The edges of the motifs are hard and do not merge with the background.

Belief 3a: The colors of deities are prescribed by religious texts. *Kṛṣṇa* and *Viṣṇu* are painted blue, *Gaṇeśa* is red, *Sarasvatī* is white, and *Kali* is black.

Belief 3b: *Ālthā* on the hands and feet is red in color.

Belief 3c: The *Bindi* on a woman's forehead is red in color, signifying her marital status.

Convention 4: The *Bāṇārasīpan* or *Bāṇārasī*-ness of the *Khilōnā* lies in its visual representation through painting and not in its shape. Therefore, shape is more flexible to change than visual representation.

Canon 4a: The higher the number of motifs filling the body space of the *Khilōnā*, the greater the skill and quality of the work. Often, the number of blank spaces on a *Khilōnā* is few. The space is filled up with *Bindi kā kām* (a visual technique consisting of dots to fill a blank space with motifs without much effort), round body patches, borders, ornamentation, or, more recently, texture.

Canon 4b: The small detailed features of a body, such as eyes, moustache, lips, fingers, toenails, and ornamentation, are painted and not carved. This is largely true for assembled human figures that share similar body parts but are differentiated through the details of the painting.

Canon 4c: Toys can be left *Plain*, displaying only their wooden texture, for modern consumption alone, whereas the shape and form of the *Khilōnā* display exceptional skill.

Canon 4d: Borders are made on most edges; for example, the *Pallu* of a *Saree* or stole, or the sides of a wooden base.

Canon 4e: The visual features are realistic. Features can be stylized but not exaggerated into a caricature.

Belief 4a: Unpainted objects are generally not found in *Bāṇārasī* consumption.

Convention 5: Facial expressions of the *Khilōnā* are impassive, without any cues to the emotional context of the story or character. Divine characters, common folk, and animals alike hold a similar neutral expression.

Canon 5a: Exaggerated caricature bodily features are neither painted nor shaped on wood.

3.1.2 Artefact: Shape Features

Convention 6: The shape of an artefact can be willfully borrowed from another *Bāṇārasī* or modern practice (unlike visual representation), especially in the absence of a similar instance of shape in the *Bāṇārasī* toy design community.

- The shape of a *Mūrti* of a deity designed for *Bāṇārasī* patrons will follow traditional precedents or those illustrated/written about in older texts. It may be simplified to reduce cost, but not modified into a new shape altogether.
- *Decoration kā Khilōnās* can freely retain or borrow shapes for toys designed for *Bāṇārasī* patrons or modern customers from within and outside the community repertoire of artefacts.

Canon 6a: Shape can be more stylized or abstract in *Khelne* and *Decoration kā Khilōnās* than painted features, which are often realistic. For modern consumption, shapes can also be made highly realistic.

Canon 6b: A base or platform is largely made for all *Bāṇārasī Khilōnās* to be placed on.

Canon 6c: The body posture of animals and human figures in contemporary *Khilōnās* will be vertical or horizontal to the ground. Diagonal or oblique postures are rare.

Belief 6a: Shapes of worldview-held sensitive features, such as *Gadā* (a club), *Śaṅkha* (conch), *Kamal* (lotus), or *Chakra* (wheel), are held constant.

Convention 7: All parts of the body, to the extent possible, will be shown in the *Khilōnā*.

- The body parts of gods, goddesses, and their *Vāhana* are determined by the worldview. They can also have multiple body parts in the case of mythical characters.
- All parts of the body of an animal or a common man are depicted.

Canon 7a: In toy designs where all body parts are not depicted, they will be willfully added.

Belief 7a: Toys should visually replicate their corresponding real objects or beings.

Belief 7b: A decapitated body is not acceptable in the worldview unless it is part of a mythological character or story.

Belief 7c: Hollow or broken *Mūrtis* are inauspicious.

3.1.3 Artefact: Size Features

Convention 8: The smaller the size of the *Khilōnā*, the lower the amount of detailing and realism.

Canon 8a: *Khilōnās* are generally of a size that can be easily handled by a single hand. Most *Khelne kā Khilōnās* fit into the palm of a hand. *Mūrtis* can be bigger but not higher than 7 or 8 inches. In the past, there have been instances of larger *Mūrtis*, but that is not the case today.

Canon 8b: *Khelne kā Khilōnās* and *Mūrtis* do not show a difference between the sizes of animals; therefore, the tiger and the elephant are of the same height. However, size variation is shown between an animal and a man.

Canon 8c: Often, a decrease in size leads to a lesser amount of detailing in shape and visual features.

Convention 9: In designs made for modern consumption, the smaller the size of the *Khilōnā*, the higher the amount of detailing and realism.

Canon 9a: *Mūrtis* change their category after a size reduction.

3.2 Conventions, Canons, and Beliefs: Production

The features of production for *Bāṇārasī Khilōnās* are also governed by a relatively small number of conventions, canons, and beliefs—which give toy designers considerable freedom to experiment with manufacturing processes, mechanisms, and materials. However, the skills available, including *Gadhwā* (carving), *Pattern* (designing), and *Kharād* (lathe work), are crucial in determining what can be achieved during production. The use of machinery and tools like the circular saw and lathe, combined with the primary material of wood, dictates how a feature can be emulated or modified during production. Consequently, while toy designers enjoy some creative freedom, the production process is still regulated by the practical constraints of available skills, tools, and materials, which influences how features are crafted and incorporated into the final products. The conventions, canons, and beliefs for the same are discussed below.

Convention 10: *Mūrtis* cannot be made hollow if they are being used for a religious purpose, including display in a *Jhāṅkī* for *Bāṇārasī* consumption. Old *Mūrtis* sold as antiques or newly made fakes of the same in the *Decoration kā Khilōnā* category are also solid and not hollow. But images of *Mūrtis* can be used in *Decoration kā Khilōnās* on hollow bodies for the purpose of display. *Mūrtis* can also be flattened into a 2D *pattern* for *Khilōnās* like jigsaw puzzles made for non-*Bāṇārasī* or modern consumption.

Canon 10a: *Mūrtis* are made with assembled carved body parts in wood. Smaller props, such as a lotus, may be made of wire.

Belief 10a: Hollow *Mūrtis* are considered inauspicious and sacrilege in the community.

Convention 11: Manufacturing techniques can be interchanged or changed in a part of an artefact. However, the whole artefact cannot be made using an unfamiliar modern manufacturing process.

- *Mūrtis* are always assembled with carved parts.
- *Khelne kā Khilōnās* are largely assembled with flat *patterns*.
- *Decoration kā Khilōnās* can be carved, use *patterns* assembly, or be turned on the *Khāraḍ*.

Canon 11a: A *Khilōnā* is mostly made from part to whole, which is true not only for *Patterns*n or *Khāraḍi kā Khilōnās* but also for some *Gadhwā Khilōnās* (carved and assembled). Parts of *Khilōnās* are shared with each other in different combinations to allow design variations and to meet the growing market demands. The hands, feet, and at times even faces are interchanged, where all figures are similarly proportioned. Cues for recognition are either provided through accessories such as *Triśūla* (a three-pronged spear symbolizing divinity in Hinduism) or lotus, or through motifs in paintings.

Belief 11a: Carving is largely done by *Viśvakarmas*, who claim themselves to be the descendants of the ivory carvers, whereas flat *Pattern* assembly is practiced largely by *Prajāpatis*, who were the traditional clay *Khilōnā* makers.

Convention 12: Willful innovation of a new mechanism is not attempted. However, existing or preceding mechanisms can be altered or modified.

- *Mūrtis* are generally static. Characters and objects of narrative *Mūrtis* are knock-down but not dynamic.
- Most *Khelne kā Khilōnās* are dynamic: pull–push *Khilōnās* on wheels, *Laṭṭus* (spinning tops) based on angular momentum, or the *Dānevālī chiṛiyā* (indigenous pecking toy) based on the pendulum.
- *Decoration kā Khilōnās* can be static or dynamic. Flying *Khilōnās* can be hung, nested dolls can be nested, rocking horses can rock, and so on.

Belief 12a: Figures of gods cannot be hung. They should be kept on a firm surface without movement.

Belief 12b: *Mūrtis* should be placed on a stable surface with no movement, not only because their allotted place in the *Jhāṅkī* or otherwise is a place of ritual that is respected, if not worshipped, but also because moving parts encourage touching, fiddling, and so on, which should be avoided to maintain their sanctity. Therefore, *Mūrtis* of deities can be enthroned on a wheeled *Ratha* (carriage, often drawn by horses), but the *Ratha* cannot be moved.

Convention 13: The material of the *Bāṇārasī Khilōnās* will not be willfully changed or replaced by another material because the skill of manufacturing a cultural design is based on the nature of the material. Generally, *Mūrtis* are assembled with carved wooden parts, and *Khilōnās* are assembled with flat wooden patterns. *Decoration kā Khilōnās* can be turned on the lathe or assembled through carved or flat wooden parts. Smaller parts, such as a tail, tongue, or parts of a crown, can be made of metal wires.

Canon 13a: The soft, *Gūlar* wood was used in the past but is increasingly becoming unavailable. *Ambva* wood has replaced *Gūlar*, even though it is hard and does not give the same finish. *Plain Khilōnās* for modern consumption are made in *Sagwan* or, if available, *Gūlar*.

Belief 13a: *Mūrtis* that are hollow are considered inauspicious because they symbolize incompleteness. *Mūrtis* should be hand-carved in wood.

3.3 Conventions, Canons, and Beliefs: Consumption

Surprisingly, just two features of consumption—purpose and connotation—together hold the largest number of conventions, canons, and beliefs in the *Bāṇārasī* worldview. The most common within this category are the beliefs, which is understandable because consumption is governed not only by aesthetics but also by beliefs that determine *Bāṇārasī* preferences and purchase decisions.

The purpose and connotation of a *Khilōnā* are strongly influenced by beliefs, including myths, taboos, and traditions surrounding the making and use of the *Khilōnās*. These beliefs shape the way *Khilōnās* are perceived and utilized within their cultural context, highlighting the intricate connection between their function and cultural significance.

3.3.1 Consumption: Purpose

Convention 14: *Khilōnās* of the *Mūrti* category can shift into another category but not vice versa. *Khelne kā Khilōnās* and *Decoration kā Khilōnās* are flexible categories that can borrow or lend their designs to other categories.

Belief 14a: The *Mūrti* category is tightly bound to the worldview because it is largely related to religion.

Belief 14b: Most *Khilōnās* of the *Mūrti* category are treated with respect and, at times, reverence.

Convention 15: Modern features are unacceptable to the *Bāṇārasī* worldview. The toy designers make them with precision and finish but do not emulate them for *Bāṇārasī* consumers. *Khilōnās* in the *Mūrti* category do not accept modern features.

Over time, through sustained client orders of modern toys, some features or designs can be accepted in the *Khelne kā Khilōnā* category.

Decoration kā Khilōnās are open to modern features, but they may not necessarily be retained.

Canon 15a: Unfamiliar modern features become visibly discordant in a *Bāṇārasī Khilōnās* or, over time, are assimilated toward *Bāṇārasīpan.*

Canon 15b: The shape of a *Khilōnā* is more malleable toward accepting modern features than the painted visual representations.

Canon 15c: *Decoration kā Khilōnās* and other modern artefacts are generally more expensive than *Mūrtis* or *Khelne kā Khilōnās.*

Belief 15a: The community respects and at times fears the worldview.

Belief 15b: The acceptability of a *Khilōnā* within the practice is more important than its innovation for its willful retention and emulation.

Belief 15c: Figurines of Christ and other Christmas decorations can be made in *Khilōnās*, but such instances are not often sold or bought in the local markets.

Belief 15d: Expensive *Decoration kā Khilōnās* and modern artefacts are not exhibited on the display shelves of the local shops. They are largely stored in a back room or a separate shop. The accessibility of modern designs is limited to certain clients alone. Beharilal Agarwal has different shops for *Bāṇārasī* patrons and non-*Bāṇārasī* customers. Similarly, another shopkeeper, Neeraj, invites selected clients to his house where the "antiques" are stored.

3.3.2 Consumption: Connotation

Convention 16: The status of a figure or animal is represented through props held by them, in addition to the worldview-held features mentioned in Convention 1.

- Ornamentation and props held by the gods and goddesses in the *Mūrti* category are largely ascertained by the religious worldview.
- The royal or divine status of the animals is often depicted by a *Chādara* on their back and body ornamentation.
- The common folk often hold props representing their occupation. For instance, the musicians hold their instruments, or the *Gwalin* holds the milk pot.

Canon 16a: The *Chādara* or cover on the back of a *Vāhan* (also, *vāhanam* typically denotes an animal or mythical entity employed as a vehicle by a Hindu deity) often has a chequered design with borders.

Canon 16b: The props of the gods and goddesses are represented realistically. Recently, golden and silver paint has been used on some.

Canon 16c: Unlike other animals, the royal elephant is depicted with body decoration.

Canon 16d: A *Nāth* or nose ring and bangles are prominently shown in female figurines.

Canon 16e: A large amount of jewellery is shown to depict divine status.

Canon 16f: Straight lines or dots of white, black, and yellow, interspersed with red, represent ornaments. Gold and silver are traditionally not used.

Belief 16a: *Viṣṇu* holds the *Gadā*, *Śaṅkha*, *Chakra*, *Kamal*, and discus in his multiple hands.

Belief 16b: *Gaṇeśa* holds the *Modak*, *Akṣamālā*, *Pādal,* a weapon, and a broken tusk in his multiple hands.

Belief 16c: *Rādhā* holds the *Pādal* or *Kamal* in her hand.

Belief 16d: *Gamchā* (a light small towel, often a hand towel) is carried by most *Bāṇārasī* men on their left shoulder.

Belief 16e: A *Chādār* is put on the back of a royal animal/vehicle or *Vāhan*.

Belief 16f: Saddlery is used on *Vāhans* to depict their status as animals used for transportation.

Convention 17: In *Bāṇārasī Khilōnās*, meaning is derived from the worldview itself. Even in instances of assimilation with modern artefacts or those originating in other cultures that are acceptable to the community, the new designs will derive their meaning from within the *Bāṇārasī* worldview.

- New meanings, such as animals, stories, or characters, cannot be introduced in rigid categories like *Mūrtis* unless they are a part of the religious worldview.
- In addition to the human characters and animal figures prescribed by the worldview, the *Khelne kā Khilōnā* category can sometimes include meanings from outside. For instance, the camel has been accepted in the *Khelne kā Khilōnā* category.
- The *Decoration kā Khilōnā* category is open to all *Bāṇārasī* or meanings borrowed from other cultural practices because it is loosely bound by the worldview.
- Abstract ludic *Khilōnās*, such as *Laṭṭus*, do not have a cultural meaning but are associated with cultural play.

Canon 17a: Largely, the diversification of an artefact will include animals and characters that have already existed in the *Bāṇārasī Khilōnās*. New gods and goddesses which have not been made before but are a part of the worldview can be added to the flying *Khilōnās*. Sometimes, the worldview can extend itself to include new animals like camels and crocodiles due to external influence.

Canon 17b: Toy designers can add new meanings from the worldview in two ways: (1) by modifying an existing *Khilōnā* to create a new one, such as converting a tiger into a dog, and (2) by borrowing visual cues from popular and easily available illustrations in calendars, advertisements, and posters.

Belief 17a: *Kāśī Nāreś* is respected in *Bāṇāras* even though the monarchy has been abolished. The ceremonial procession of the *Kāśī Nāreś* riding a royal elephant is still practiced during festivals in *Bāṇāras* by the successive nominal kings and is represented in *Khilōnās*.

Belief 17b: *Kāl Bhairava*'s (a powerful manifestation, or avatar, of the Hindu God Śiva) vehicle is a black dog, and in some places, dogs are fed on this day. Black dogs, specifically, are considered a form of *Bhairava*. Devotion to Lord *Bhairav* can be shown by feeding and taking care of dogs every Saturday. All problems are believed to be resolved by this simple act of charity.

Belief 17c: The elephant, tiger, peacock, parrot, and cow are the five favourite animals of *Thākurjī*.

Belief 17d: *Jhāṅkīs* largely consist of stories from Lord *Kṛṣṇa's* life.

Belief 17e: The relationship between people, animals, and objects is fixed in the worldview. Therefore, gods and their escorts, *Vāhanas*, and props are fixed as well. For example, *Śiva* is depicted with his escort *Parvati* and *Kṛṣṇa* with *Rādhā*.

Belief 17f: Animals and characters from *Pañcatantra* and other folk stories are not included in the *Bāṇārasī Khilōnās*, even though they are familiar to the community.

Belief 17e: Figures of Christ and other Christian artefacts can be made in the community but not sold or bought by the community.

Convention 18: For *Bāṇārasī* patrons, meaning and category are related; when categories change, meaning also changes, where meaning largely refers to purpose.

- *Mūrtis* can shift categories but do not allow artefacts from other categories to shift into their own. They change their purpose when they shift.
- *Khelne kā Khilōnās* and *Decoration kā Khilōnās* borrow and lend members between categories, provided they fit their purpose of play or display, respectively.

Canon 18a: In the process of assimilation, popular familiar images are freely borrowed by toy designers from books, magazines, and calendars to assist the process of change with a non-*Bāṇārasī* or modern artefact.

Canon 18b: *Khilōnās* designed for modern consumers with a sustained client demand is likely to be assimilated into a community-held category with a *Bāṇārasī* meaning.

Canon 18c: Sacred images can be stripped of their ritualistic significance and sold for display purposes alone to modern customers.

Belief 18a: Ritualistic and religious meanings are respected and closely adhered to by the *Bāṇārasī* community.

Belief 18b: Largely, *Bāṇārasī Khilōnās* play an enculturative role in the community; therefore, the values and meanings represented by the *Khilōnās* are significant.

Convention 19: Discordant features can become a part of the *Bāṇārasī* identity in the absence of a suitable replacement or if frequently emulated in the community.

Canon 19a: Discordant features are borrowed from modern and unfamiliar practices.

Canon 19b: Normally, discordant features will be omitted from a *Khilōnā* after the client order has been fulfilled.

Belief 19a: The worldview is not open to discordant features in the *Mūrti* category.

Belief 19b: Discordant shapes such as multiple arms and heads are sanctioned for particular deities by the worldview.

Belief 19c: Emulation is an unauthored method of community transfer, retention, and evaluation of knowledge.

Convention 20: Only a single frame of the climax of a mythical story is depicted.

- Narrative compositions are depicted in *Mūrtis* alone.

- *Khelne kā Khilōnās* and *Decoration kā Khilōnās* may have a mythical character but not a story.

Canon 20a: Narrative *Mūrtis* are knock-down compositions.

Canon 20b: The knock-down system enables portability and easy storage of large heavy compositions.

Belief 20a: Stories for the *Jhāṅkīs* are borrowed from the life of Kṛṣṇa.

Belief 20b: *Nāga-nāthīya* is a festival in *Bāṇāras* in which a child is dressed up as Kṛṣṇa. The child Kṛṣṇa dances on the head of a wooden snake in the river Ganga.

Belief 20c: Narrative compositions are made seasonally during *Janmāṣṭamī*.

Belief 20d: The wooden compositions designed for *Jhāṅkīs* are wrapped in cloth and preserved for the next years.

Belief 20e: The family along with the children display the compositions in a *Jhāṅkī*, which is removed on the *Chati* (the sixth day after birth).

Belief 20f: If the *Mūrti* of the god in the composition breaks, it is offered to the river Ganga and not disposed of in a dustbin.

Convention 21: Taboos can be broken in non-*Bāṇārasī* or modern designs.

- Figures of gods and goddesses and their *Vāhanas* rigidly follow the worldview.
- *Khilōnās* often borrow the meaning from the worldview but are also open to other familiar meanings from different situations.
- *Flying Khilōnās* and nested dolls break a taboo but follow all the visual features and props associated with the deities. Taboos are easily broken in client orders.

Canon 21a: The taboos are not broken willfully by the toy designers unless the artefact is made for non-*Bāṇārasī* or modern consumers.

Canon 21b: Taboos can be broken for non-*Bāṇārasī* or modern consumers who do not understand or are indifferent to their significance.

Belief 21a: Owl or *Ulukā* becomes the vehicle of Goddess *Lakṣmī* when wealth and prosperity provided by Goddess *Lakṣmī* is used by human beings for *Adharmic* or corrupt actions. Symbolically, the wealth provided by *Lakṣmī* blinds the person and leads to dishonesty and avarice. Therefore, the owl is considered a bad omen.

Belief 21b: The figurines of gods should not be left hanging in mid-air.

Belief 21c: Artefacts that are tabooed in the worldview can be made for non-*Bāṇārasī* or modern consumption but cannot be sold in the *Bāṇārasī* markets.

Convention 22: Negative characters are not portrayed in the *Khilōnās*.

- Characters of gods and goddesses and their stories are depicted in the *Mūrtis* as role models for children and others. The stories may include a villain only to show his defeat at the hands of the righteous. Effigy of a *Rāvaṇa* is made during *Rāmlīlā* only to be burned down at the end of the ten-day festival.

- In some *Khāraḍi kā Khilōnās*, common *Bāṇārasī* folk are also depicted but without any social or moral vice. No ordinary *Bāṇārasī* resident is depicted engaging in drinking, gambling, or playing cards—activities that are disapproved of in *Bāṇārasī* culture.
- *Khelne kā Khilōnās* largely include animal characters that are the *Vāhanas* of the deities in the *Mūrti* category. No evil animal characters are depicted.
- *Decoration kā Khilōnā* may include a character like Ravan but only for non-*Bāṇārasī* or modern consumption.

Canon 22a: Skeletons are not represented in the *Bāṇārasī Khilōnās.*

Canon 22b: The body image of a doll should not evoke feelings of desire.

Belief 22a: In *Bāṇāras*, consuming alcohol and gambling are mortal sins.

Belief 22b: Ravan is not worshipped because he was a *Rākṣasa* (devil).

Belief 22c: *Khilōnās* impart values and culturally upheld morals to children. Stories of Kṛṣṇa and other deities help children learn about virtuous actions and behavior and instill respect for religious principles.

Convention 23: All human figures are healthy young men and women. Aged characters are not depicted. The only instance is that of Lord *Brahmā*, who is mythically considered to be an old man.

Canon 23a: The body of men is depicted as healthy but never with muscles.

Canon 23b: Lord *Brahmā* is clad in red clothes. He is traditionally depicted with four heads, four faces, four arms, and a white beard. In his hands, he holds a sceptre, a book, a string of prayer beads called the *Akṣamālā* (literally "garland of eyes"), and the *Vedas*.

Belief 23a: *Brahmā* is often referred to as the grandsire of all human beings.

The extensive set of regulations that governs the artefacts and their production and consumption reveals a clear predominance of beliefs, which tally at 55, followed closely by canons at 49, and conventions at 22. Not only are beliefs more numerous, but they also demonstrate a remarkable resistance to change, especially when contrasted with canons and conventions. An artefact or a *Khilōnā*—along with its production and consumption—is significantly shaped by beliefs associated with its purpose (e.g., display in a *Jhāṅkī*) and its connotation (e.g., *Kṛṣṇa* as a deity whose birth is celebrated through the *Jhāṅkī*) rather than by its aesthetic value. As a result, canons and conventions serve to articulate these beliefs. For example, Belief 20b recounts the tale of *Nāga-nāthīya*, where the child *Kṛṣṇa* overcomes a massive sea serpent and dances upon its head. This belief is reflected in Convention 20, which stipulates that a single frame capturing the climax of a narrative is to be depicted in *Mūrtis*. Furthermore, Canons 20a and 20b pertain to the modular representation of

Fig. 5 Wooden depiction of *Nāga-nāthīya* in which all characters can be disassembled from the base. *Image source* Author

large compositions. These conventions, canons, and beliefs can be clearly seen in Fig. 5.

Unlike conventions and canons, which emerge specifically in relation to a *Khilōnā*, beliefs are derived from a wider array of social and religious practices. This broad source base includes a multitude of myths, taboos, and traditions, thereby extending the influence of beliefs over a variety of community activities. This vast reservoir accounts for a substantial number of beliefs. The enculturation of the *Bāṇārasī Khilōnā* is predominantly a manifestation of these beliefs through its conventions and canons of thinking, making, and using *Khilōnās*. *Bāṇārasī* toy designers and patrons only accept *Khilōnās* that conform to these conventions, canons, and beliefs, thereby determining the appropriateness of a *Khilōnā* within the *Bāṇārasī* worldview. For example, the *Bāṇārasī* concept of असभ्य(*asabhya*) refers to representations deemed indecent or uncultured and therefore unfit within this worldview. Most parents in *Bāṇāras* adhere to the convention that a curvaceous doll with make-up is inappropriate for a young girl. Barbie dolls, which exhibit bodily curves, are considered *asabhya* because they invoke sexual desire. The appropriate doll for a young girl

Fig. 6 Flying *Khilōnā* with *Gaṇeśa* and bird. *Image source* Author

represents *Rādhā*, the childhood friend and lover of Lord Kṛṣṇa (Convention 22, Canons 1c, 1e, and 22b, Beliefs 1c, 3c, and 22c).

Historically, the authority to determine the propriety of such features rested with the kings or the *Purohits*. However, today, this power lies solely with the community of producers and consumers. The acceptability of a new design is assessed either within the *Karkhanā* through the toy designer's judgment or by consumers in the market. For instance, Ajit Kumar Viśvakarma, a toy designer, attempted to innovate by changing Lord *Gaṇeśa*'s vehicle from a rat to a peacock. This deviation from the belief was quickly met with criticism from both local patrons and foreigners, who refused to purchase the *Khilōnā*. Consequently, Ajit was forced to withdraw the *Khilōnā* from the market (refer to Fig. 6).

Thus, the typicality of a *Bāṇārasī Khilōnā* arises from a nuanced interplay of apprenticeship, communal participation, ritual practices, oral traditions, and market dynamics. These interconnected elements work together to ensure that the rules guiding the creation of *Khilōnās* are not only learned and shared but also perpetuated across generations, thereby preserving the integrity and distinctiveness of this generational practice of cultural design.

References

Alexander C (1964) Notes on the synthesis of form. Harvard University Press

Björklund L (2009) The forming and assessment of creative skills, from a neurocognitive point of view. In: PATT-22 conference, Delft, The Netherlands, 24–28 Aug 2009

Coomaraswamy AK (1977) Art of Eastern Asia. In: Lipsey R (ed) Coomarswamy: selected papers. Princeton University Press, New Jersey, pp 80–113

Cross N (1982) Designerly ways of knowing. Des Stud 3(4):221–227

Dundes A (1989) Folklore matters. The University of Tennessee Press, Knoxville

Katz D (1960) The functional approach to the study of attitudes. Public Opin Q 24(2):163–204

Lee K-P (2004) Design methods for cross-cultural collaborative design project. In: Design research society international conference: Futureground. Monash University, Melbourne, p 135

Leong BD, Clark H (2003) Culture-based knowledge towards new design thinking and practice: a dialogue. Des Issues 19(3):48–58

Leung K, Bond MH, de Carrasquel SR, Muñoz C, Hernández M, Murakami F, Yamaguchi S, Bierbrauer G, Singelis TM (2002) Social axioms: the search for universal dimensions of general beliefs about how the world functions. J Cross-cult Psychol 33(3):286–302

Lin R-T (2007) Transforming Taiwan aboriginal cultural features into modern product design: a case study of a cross-cultural product design model. Int J des 1(2):45–53

Millikan RG (2005) Language: a biological model. Oxford University Press

Moalosi R, Popovic V, Hickling-Hudson A (2010) Culture-orientated product design. Int J Technol des Educ 20:175–190

Norman D (1999) Affordance, conventions and design. Interactions 38–43

Redfield R (1941) The folk culture of Yucatan. University of Chicago Press, Chicago

Chapter 5
Hand Over Mind: Pedagogical Notions of the "Others"

> What one acquires here is not a technique; one learns correct judgements. There are also rules, but they do not form a system and only experienced people can apply them right, unlike calculating rules. (Wittgenstein 1963, 227)

1 The Taste of Colour

In his essay "The Implicit and the Manifest in Indian Folk Art and Mythology," Jain (1993) offers a poignant contrast between the urban literate traditions of creative history, as enshrined in canons and principles like the *Śilpaśāstra*, and the living rural traditions embodied by the ordinary rural toy designer. Jain recounts a compelling interaction between a well-trained ethnographer and Khatri Mohammad Siddik, a dyer and printer from the village of Dhamadka in Kutch, Gujarat.

The ethnographer, steeped in a tradition of precise measurements, identifiable proportions, and explicit know-how, sought to understand the exact quantity of alum required to mix with mud to prepare a solution of resist-mordant material. The intensity of the colour, he understood, would depend on the alum's proportion—the higher the proportion, the deeper the colour. However, Khatri Mohammad Siddik, rooted in the tacit knowledge of his cultural design practice, had no specific weights or proportions to offer the erudite ethnographer.

After an extended session of descriptive, informal questioning, the ethnographer, frustrated by the lack of concrete data, requested Khatri to demonstrate his method of mixing alum. Khatri obliged by dipping his fingers into the mud and alum solution and rolling the mixture on his tongue for a moment. Dissatisfied, he added a bit more

---

This chapter draws upon material from the author's earlier publication, "*Inexplicit Learning: Transferring Knowledge through Visual and Emulative Practices*," in Techne Serien—Forskning i slöjdpedagogik och slöjdvetenskap, 24(2), 2017. The current version has been revised for clarity and contextualisation.

K. Patil, *Toying with the Mind*, Design Science and Innovation,
https://doi.org/10.1007/978-981-96-9227-9_5

alum, tasted the mixture again, and finally, satisfied, declared the concoction ready to produce a deep red, akin to a cloth lying nearby.

In Khatri's words: The scale can fail you, but how can your tongue fail you? When you drink your tea or eat your food, don't you know whether there is too much sugar or too little salt in it? I can produce five different buckets of mordant-cum-resist material on five different days and dye five different pieces of cloth in them by my method of judging the proportion of alum, and all five cloth pieces will yield exactly the same shade of red. As I savour alum on my tongue, my eyes visualise the corresponding shade of red that I will get. The son of a *Khatri* (dyer) can never get a girl in marriage until he passes this test of judging the right proportion of ingredients by using his senses and not gadgets.

Jain's narrative highlights the profound and nuanced understanding inherent in cultural designs, where sensory experience and tacit knowledge transcend the empirical precision of modern methodologies. This anecdote illustrates the deep-seated, embodied knowledge that rural toy designers possess, which is often undervalued or misunderstood by those outside their cultural and practical milieu.

Several aspects of this anecdote are intriguing. The most obvious one could be—how can taste be an identifier of colour? Khatri Mohammad Siddik appeared to visualize colour as he savoured the mixture of mud and alum on his tongue. The correlation between these two senses is not well understood, which makes the transfer of this knowledge to the ethnographer, and even to us, very challenging. Not only is the knowledge tacit, but it also involves a lesser-known interrelationship between our senses. However, this knowledge and skill seem to transfer effortlessly between the master and the apprentice in cultural design communities.

Learning a skill that involves bodily actions reflecting the mind requires not only manual dexterity but also body–mind coordination facilitated by sensory stimuli. The design process unfolds inside the designer's head, and perhaps body, as well as outside in the observable world (Reitan 2014, 3). Such tacit knowledge is not easily conveyed through short courses or non-apprenticeship modes like distance education (Muller 2000). This underscores a significant divide between teaching theoretical concepts and hand-based thinking skills. The latter demands the regulation, manipulation, and synchronization of many more senses than the mind alone. This makes a pedagogical inquiry centered around the transfer and transformation of such skills and designer knowledge difficult to examine empirically due to its intangible or inexplicit nature.

The apprenticeship tradition has been the primary source of vocational training in cultural design communities for centuries (Wolf 2002, 58). Polanyi (1958) elucidates the apprenticeship relationship as one where the master controls the theoretical, generative, and evaluative principles of the practice, not the novice. Brown et al. (1989) explain that cultural design apprenticeship enables apprentices to acquire and develop the tools and skills of their trade through authentic work and membership in their community. Thus, the term *apprenticeship* emphasizes the centrality of activity in learning and knowledge building and highlights the inherently context-dependent, situated, and enculturating nature of learning. Rogoff (1994, 213) explicates that less mature learners in a community learn through collaboration with adults while engaging in activities that are explicitly connected with the community's history and current practices. This curricular mode, which was mainstream before the Industrial Revolution, is now considered an alternative system. In modern workplaces, such learning is often treated as a residual category that describes any kind of learning that does not take place within or follow from a formally organized learning program or event (Eraut 2000, 114).

Nevertheless, cultural design apprenticeship remains a rich curriculum resource, not despite but because of the current emphasis on flexible specialization. In this context, knowledge practices are tacit, extensive, distributed, and dispersed; develop within a coordinated network of people, tools, and technologies; and serve multiple, integrated, and overlapping functions (Gamble 2004). Thus, concepts of intuitive expertise and tacit knowledge have become increasingly important in discussions concerning knowledge and learning in the practice of skill acquisition and its application (Nielsen 2002).

Moreover, the tacit knowledge accrued through long practice in cultural design and the instrumental practice of "doing the job well for its own sake" (Sennett 2008, 9) suggests new ways of doing things (Muller 2012). This implicit expertise, cultivated over years of dedicated practice, reveals innovative approaches that might not emerge from formal, theoretical instruction alone.

In cultural design communities, the apprenticeship model ensures that this deep, embodied knowledge is transmitted seamlessly from one generation to the next. Through hands-on learning and direct engagement with experienced masters, apprentices absorb not only the technical skills but also the cultural and aesthetic sensibilities that define their cultural design practice. This method stands in stark contrast to the modern educational emphasis on explicit knowledge transfer and highlights the importance of context, experience, and embodied practice in true skill acquisition.

This approach underscores the value of learning environments that integrate sensory engagement and practical application. The apprenticeship model's effectiveness in fostering such environments suggests potential pathways for modern educational practices to embrace more holistic, context-dependent learning strategies. By acknowledging and incorporating the nuanced, tacit dimensions of knowledge, contemporary education can better prepare individuals for the complexities and subtleties of real-world applications in various fields.

According to Nonaka and Takeuchi (1995), knowledge is created through the interaction between tacit and explicit practices, which postulates four different modes of knowledge conversion:

Tacit-to-Explicit: This is the process of externalization, where tacit knowledge is articulated into explicit concepts. It involves converting personal, implicit knowledge into a form that can be shared and understood by others.

Explicit-to-Tacit: This mode, known as internalization, involves the absorption of explicit knowledge into tacit knowledge. It is the process of learning and integrating formal knowledge into one's own practices and experiences, making it part of one's implicit understanding.

Explicit-to-Explicit: This is the combination process, where different explicit knowledge elements are systematized into a cohesive knowledge system. It involves synthesizing and integrating various pieces of explicit information to create a structured and organized body of knowledge.

Tacit-to-Tacit: Known as socialization, this mode involves sharing experiences to create tacit knowledge, such as shared mental models and technical skills. It is the process by which individuals gain tacit knowledge through direct interaction and shared experiences, often without formal instruction.

These modes of knowledge conversion illustrate the dynamic and cyclical nature of knowledge creation and sharing within organizations and communities (refer to Fig. 1). Each mode highlights the importance of both tacit and explicit knowledge and the continuous interplay between them in developing and sustaining a robust knowledge base.

Pedagogical methods for knowledge conversion from and to explicit practices are fairly well known. However, from Nonaka's perspective, we must inquire into the pedagogical methods that enable the completion of the knowledge spiral in cultural

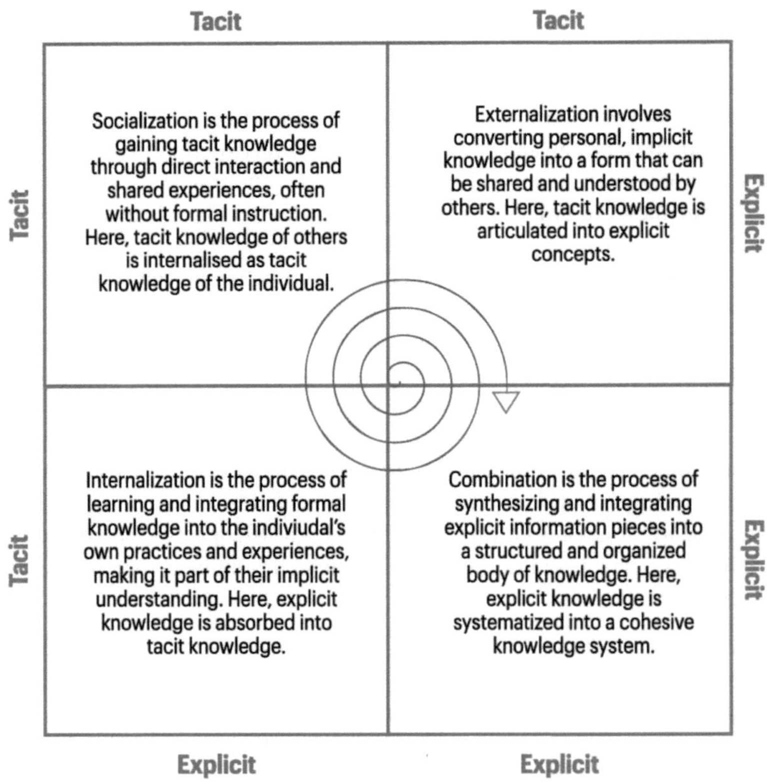

Fig. 1 The knowledge spiral proposed by Nonaka and Takeuchi (1995). *Image source* Recreated by the author from the knowledge spiral proposed by Nonaka and Takeuchi (1995). It was also published by the author in Patil (2025)

design communities. Studies on methods of transfer and acquisition of tacit knowledge are limited, often speculative, and rarely generative. The "tacit" has become a stumbling block, both practically and conceptually (Gamble 2004). Little is understood about how, in the absence or lack of any explicit communication, the master hands down his knowledge to the apprentice, and in turn, the apprentice imbibes this tacit knowledge.

The central concern of this chapter is to develop a body of knowledge to understand how expert knowledge is transmitted in the absence or lack of explicit and formal means of knowledge acquisition. I will particularly focus on the evolved pedagogical practices of cultural design communities that center around a master–apprentice relationship.

2 The Classroom and the *Karkhanā*

Do you remember Newton's three laws of motion? Across various curricula worldwide, Newton's three laws of motion are often introduced in grade 8 or 9 physics textbooks. Do you recall how these laws were taught to you in school? My textbook initially introduced the concept of force, followed by its measurement unit, and then the three laws themselves. My school teacher posed a question after teaching the laws: If a ball and a feather are dropped simultaneously from the same height, which will hit the ground first? Although we did not experiment, we hypothesized the outcome. The textbook chapter then provided everyday examples to illustrate the three laws of motion. The pedagogical principle at work in my formal education was to progress from a general law to specific instances of its application. Thus, the learning trajectory moved from general to particular, and from a theoretical paradigm to practical application.

Interestingly, much later in life, in a similar formal educational context of higher education, I learned to operate a lathe machine in the workshop by following a similar format, albeit for a hand-based skill. The lab instructor adhered to several steps:

1. *Introduction*: The instructor introduced the function of all parts of the lathe machine.
2. *Demonstration*: The instructor demonstrated the simple turning of a cylinder on the machine and provided a verbal explanation.
3. *Imitation*: I was then asked to follow his lead and turn a cylinder on the machine.
4. *Repetition with complexity*: Steps 2 and 3 were repeated with more complex mathematical proportions.
5. *Practice and correction*: The instructor asked me to practice turning the cylinder and other complex shapes, and he corrected my mistakes periodically.
6. *Application*: Once the instructor was satisfied with my skill in creating abstract shapes, he asked me to turn the wood to make a pencil stand.

Here too, the pedagogical principle at work is the same as how Newton's law were taught to Newton's laws were taught to me at school. To operate the lathe machine, first, the instructor taught me how to make a general or abstract shape, like a cylinder. He then introduced complex shapes that were also not meaningful or useful in any sense. Lastly, the instructor asked me to make a particular or specific shape like that of a pencil stand. Thus, the learning trajectory moved from general to specific and from theory to practice.

However, in an ethnographic study, I observed *Bāṇārasī* master toy designers teach their apprentices differently. For weeks and months, the apprentice was only asked to observe the master while doing odd jobs around the workshop, such as bringing the master a cup of tea. From my field notes, I was able to draw out the following teaching–learning steps followed in the workshop:

1. *Observation*: The apprentices observe the master for weeks and months, with minimal and curt verbal interactions.
2. *Unskilled work*: The apprentices complete minor unskilled tasks, such as polishing a toy made by the master with emery paper.
3. *First task*: The apprentices make a spinning top. The master quickly reprimands for errors and signals nods of approval for good attempts.
4. *Repetition*: The apprentices repeat the process of making spinning tops multiple times, experimenting with various sizes and shapes.
5. *Variety*: The apprentices continue to repeat the previous steps to make a greater variety of *Khilōnās*.
6. *Evaluation*: The master evaluates the apprentices from time to time with tasks and technical questions to ascertain their learning curve.

In contrast to the structured environments of physics classrooms or technical workshops found within formal education systems, the *Bāṇārasī* toy designers offer a compelling example of informal learning. Within this community, apprentices embark on a learning journey that begins with a focus on a specific object, such as a spinning top, and gradually expands to encompass a broader understanding of the turning process as a manufacturing technique. Here, theoretical principles emerge organically from practice rather than being imposed on it.

A similar pattern emerges among the seamstresses of Kaktovik, an Inuit village in Alaska, where novice seamstresses are expected to create an entire *atigi*—a traditional parka—on their first try, without prior practice on its parts. Reitan (2006), an ethnographer, observed no instances of young seamstresses being tasked with making sections of an *atigi* for practice. Instead, they learned the intricacies of the whole garment from the start, guided only by the opportunity to observe and imitate the work of adults.

These examples suggest that the learning trajectory progresses from the particular to the general in informal, hand-dominated cultural industries. Here, the particular is an object integral to practice, and the general is the comprehensive understanding or theoretical framework that supports the practice. This prompts an intriguing inquiry: What type of learning objects serve as the particular through which apprentices acquire skills in this context? Can any object fulfil this role? According to L'Allier (1997), a learning object is the smallest independent structural experience that includes a well-aligned objective, a learning activity, and an assessment that effectively evaluates the stated objective.

In the *Bāṇārasī* tradition of toy making, apprentices in the *Karkhanā* are often tasked with creating a spinning top (in the lathe turners workshop), a *Hilantā* (in the *pattern* makers workshop), or a wooden cow (in the carving workshop). Upon close examination, these objects possess several attributes that make them particularly suitable as learning objects, more so than other *Khilōnās*.

First, these *Khilōnās* serve as prototypical examples within their category, not only because of their typicality but also due to their perceived or actual value to the community. In her seminal paper on natural categories, Rosch (1973) argues that certain stimuli are better exemplars of a concept than others and that natural categories are easier to learn. According to prototype theory, a prototype is the central tendency of a category and represents the most typical example of a concept by (a) sharing the maximum number of features or attributes, and (b) having a minimum overlap with instances of other concepts. These prototypical examples are not only easily identifiable but also more readily learned than other members of their category.

In *Bāṇāras*, spinning tops and pecking *Khilōnās* are designed more frequently than most other *Khilōnās* and are among the oldest designs still in production. These *Khilōnās* are familiar to both young and old, as well as to consumers beyond *Bāṇāras*. Their frequent presence has cemented them as prototypical examples of *Bāṇārasī Khilōnās*. Instead of introducing a highly innovative and obscure design, master toy designers choose to impart the skill of toy making through these common and well-known examples.

Second, these prototypical *Khilōnās* are typically small and feature large, plain surfaces that do not demand highly refined motor skills. This characteristic encourages experimentation since the pieces require minimal raw materials, thereby reducing costs associated with trial and error. The *Khilōnās* are inexpensive, and their production requires less time, which leads to a quick turnaround. The process is streamlined and well understood within the community and thus minimises the risk of failure. Even if a failure occurs, the low production cost means it does not significantly impact the economic well-being of the *Karkhanā*.

Third, these *Khilōnās* enjoy consistent popularity among consumers and are in demand year-round. Unlike seasonal *Khilōnās*, these prototypical examples are always sought after, which ensures steady sales. The demand for the products also provides an opportunity for toy makers to discreetly sell *Khilōnās* with minor flaws made by apprentices to less discerning customers, thereby allowing apprentices to contribute to the family's or *Karkhanā*'s income as soon as they master the cultural design practice.

Finally, these *Khilōnās* are easily identified as representative of the wooden toy designers' cultural design. Many consumers, knowledgeable about cultural design traditions, can readily associate these prototypical examples with *Bāṇārasī* generational practices. Their ease of recall further underscores the exemplary status these *Khilōnās* hold, which the master uses to teach apprentices the skills of toy making.

These objects are integral to the "authentic" culture, as described by Brown et al. (1989). Their meaning and purpose are socially constructed through the ongoing negotiations among both present and past community members. Authentic activities enable practitioners to act with meaning and purpose while refining their tools and techniques. In hand-based learning, the master's ability to select the appropriate learning object is crucial because this choice is pivotal to achieving pedagogical aims. True to the scholarship on learning objects, these prototypical *Khilōnās* are accessible to everyone within the learning community, can be used repeatedly, and are interoperable across various studios, masters, and learners.

3 Participating to Learn Principles: Community of Participation

According to Lave and Wenger (1991), the focus should be on the types of social engagements that provide a context for learning rather than on cognitive processes and conceptual structures involved in knowledge transfer. This perspective shifts learning from the acquisition of propositional knowledge to a form of learning situated within the community's specific form of co-participation. The mode of participation, which varies across communities and situations, can illuminate the community's pedagogy in passing tacit knowledge from master to apprentice.

In the *Bāṇārasī* community of toy designers, nearly all *Khilōnās* require multiple skill sets, which fosters mutual dependence among designers to complete a product. Each toy designer specializes in a particular skill set—such as carving, lathe turning, *pattern* cutting and assembling, or painting. Community members often operate without a strict hierarchy, with no single leader directing others. For any given client order, they collaborate based on the skills required. Each group member completes their part and passes it to the next designer. If changes are made to the toy, the subsequent designer is expected to seamlessly complement the adjustments with their skill, even in the absence of explicit instructions. These skills consistently arise from the same level of technology, materials, and design for each new order. For a detailed workflow, please refer to Fig. 2.

The hierarchy-less participation in the *Bāṇārasī* toy designers' community emerges from the pedagogical methods employed within the *Karkhanā*, which are homes-cum-workshops of master toy designers. These methods necessitate engagement with in situ production activities and allow apprentices to interact with local technologies of practice, participate in culturally sanctioned interactions with the *Khilōnās*, co-create the social representation and value of *Khilōnās*, and gain opportunities to observe and listen to senior members. Moreover, they provide ample time and space to familiarize oneself with resources and the know-how of their application and use. *Karkhanās* accept the labour of apprentices, whether from within or outside the family, in exchange for the opportunity to learn the practice (Lave and Wenger 1991). They function as institutions of learning, factory floors for design and production, and living spaces for social interaction. Within a single roof,

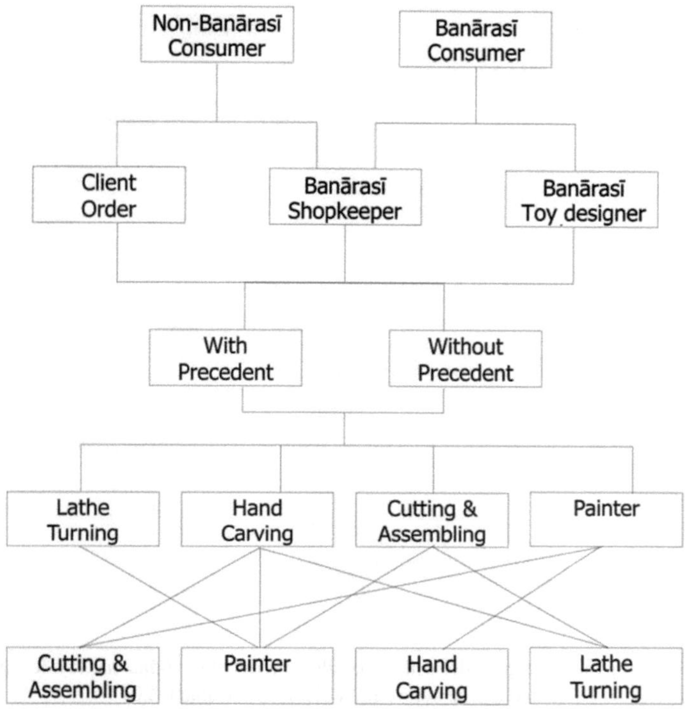

Fig. 2 Workflow of all community stakeholders involved in designing a toy (Patil 2017a, b). *Image source* Author. Also published by the author in Patil and Athavankar (2017)

multiple kinships—such as master–apprentice, employer–employee, and head of the family–family members—are commonly observed (Patil 2017a, b).

The task segregation within the aforementioned workflow is made possible by the grouping of skill sets and linguistic categories (Patil 2015), as each member performs tasks in an assembly line system and specialises in a singular, holistic, and meaningful task and skill set. Each *Karkhanā* specializes in a particular skill set and imparts related knowledge. However, in the toy-making sequence, members engage in multiple combinations of skills in varying sequences to optimize their capabilities alongside others. These skill sets have evolved and been mastered within *Karkhanās* across generations, as evidenced by kinship maps. Thus, in the workflow model discussed earlier (Fig. 1), a cutting and assembling toy designer can be easily substituted by kin, such as a father (F). Here, F is both a father and a master specializing in cutting and assembling wood for production, who also exchanges the apprentice's labour for his generational knowledge. In other words, the workflow diagram can be expanded to include relationships beyond those shared by the master alone. It can also accommodate apprentices whose labour is utilized to varying extents to fulfill client orders. This shifts the focus from the apprentice as a learner to the apprentice

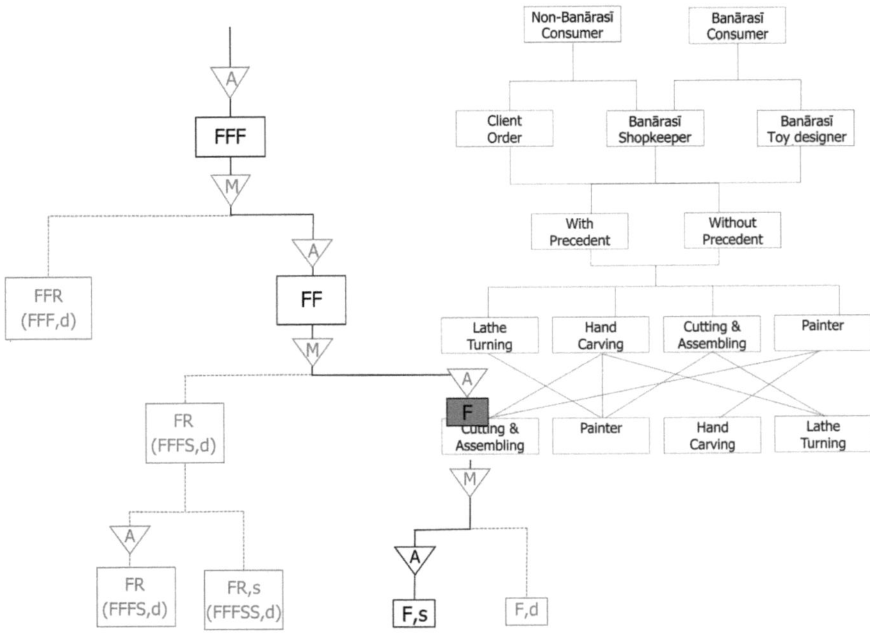

Fig. 3 Substituting the kinship model with a workflow model of the *Bāṇārasī* toy designers. *Image source* Author. Also published by the author in Patil and Athavankar (2017)

as an active participant in both production and social processes. The apprentice is a toy-designer-in-becoming rather than a toy-designer-to-become (refer to Fig. 3).

Functional specialization and segregation within the *Karkhanā* limit the transfer of knowledge between master and apprentice to what can be seen, emulated, practiced, and modified within the workshop's confines. The walls of the *Karkhanā* thus shape the curriculum's scope—determining who participates in what activities, how masters conduct themselves, how community members interact, and what apprentices must learn to fully integrate into the community. This includes developing an understanding of how, when, and on what subjects masters collaborate, collude, and sometimes clash, as well as what they enjoy, dislike, respect, and admire (Lave and Wenger 1991, 95). The *Karkhanā* serves as a microcosm that provides role models in the form of masters, advanced apprentices, and completed products.

At any given time, interactions within the *Karkhanās* involve not only the selected master–apprentice pair but also spontaneous interactions among family members and outsiders. An apprentice is not confined to learning solely from the master but is encouraged to assimilate and synthesise knowledge informally through the surrounding environment. Consequently, studying a social learning system in this context differs from experimental studies where subjects are explicitly identified, and learning interactions are limited to one-to-one communication rather than the broader, one-to-many associations prevalent in the *Karkhanā.*

Informal interviews for understanding the pedagogical modes within the *Karkhanās* were shaped by the personal, interpersonal, and community processes of interaction visible and accessible to me as a participant in community activities. These conversations naturally evolved into deeper discussions and actions that explored the relationships between master and apprentice as they lived, learned, and earned their livelihood under the same roof. With participants' consent, some interactions were video recorded (8 sessions of about 2 h each), while others were audio recorded (10 sessions of approximately 1 h 15 min each). Codes emerged from the content analysis of the transcriptions. The data were treated as a cross-dimensional comparison of verbal utterances, observed actions and processes, and recorded still photographs or videos of in situ action during the period 2008–2012. Axial patterns were inferred by connecting selective codes to develop a basic model of who teaches what to whom, by which means, and with what effect.

Additionally, this study draws inferences from previous publications on the conventions, canons, and beliefs held by toy designers; the linguistic categorization of the *Khilōnās*; and the division of skills among toy designers as discussed in Chaps. 2–4. Together, these insights inform the findings of this chapter on how toy designers transfer tacit knowledge.

4 Visual Learning

Rotuma, a hinterland community located about three hundred miles north of Fiji, was the subject of an ethnographic study conducted by Howard (1985) in the 1960s while the region was still under British colonial rule. Howard sought to understand how the modern native, caught between his own culture and British influences, struggles to maintain his identity as a Rotuman. In this community, personal demonstration was the preferred educational technique used by adults to enculturate the younger generation. It is, therefore, unsurprising that the entire learning effort was directed toward shaping children to be sensitive to role models.

As Howard observed, "Since children are so frequently in the company of adults while the latter are working, they have a great many opportunities to observe how essential tasks are performed. Children's efforts at imitation are encouraged with subtle praise or non-verbal expressions of approval, unless, of course, they start to 'show off.' If a child experiences difficulty in performing an action, an adult might physically manipulate the child's body to correct an error or refine a movement, but explicit verbal instruction is rarely offered, and children rarely ask for it. If they do, they are likely to be told to watch a skilful adult in action" (Howard 1985, 116).

Such cultural learning methods are common in communities that have not formalized the education of their children. In Kaktovik, young and adult seamstresses constantly observe each other. The design community expects a debutante seamstress to design and stitch an entire garment singlehandedly by simply watching the skilled adult seamstresses, without any assistance in the form of explicit instructions, encouragement, or even admonishment. Similarly, in the *Bāṇārasī* community of toy designers, as detailed in earlier field notes, the teaching–learning process between master and apprentice begins with observation.

The journey of becoming a skilled toy designer in the *Bāṇārasī Karkhanās* begins with an extended period of observation, where the apprentice closely watches the master at work. This phase, which can last for weeks or even months, involves minimal verbal interaction and emphasizes learning through careful observation.

From a young age, children of *Bāṇārasī* toy designers are immersed in the environment of the *Karkhanā* because it serves both as a workshop and a family home. This natural integration into the workspace allows children to start observing the toy designing skills from an early age. The significance of this observational learning is highlighted in various transcriptions from local toy designers (refer to Transcriptions 1, 2, and 3).

I watched my father and learned. Nobody can learn in a hurry. In the first one or two months, the master will allow you only to watch. Then he might give you a saw. Later, he will show you how to use a chisel.

Transcription 1: Arun Prajāpati *Karkhanā* 2, Khojwa

I learned to shape wood in *Bāṇāras*. I used to work in the house of a neighbouring toy designer. Watching the other designer shape wood, I also slowly learned how to copy the same.

Transcription 2: Ramlal Prajapati *Karkhanā* 3, Kashmiriganj

Children learn by themselves. They just come, sit, and watch.

Transcription 3: Ratanlal Viśvakarma *Karkhanā* 6, Laksha

This early exposure to the practice through observation is tied to the community mode of production, where work and home are intertwined. Gaskins and Paradise (2010) note that in societies where adults work at home, children are more likely to be involved in the activities, learning by watching rather than through formal instruction. This aligns with Bandura's (1971) theory that much of learning can occur through vicarious observation, where the observer absorbs complex behavior by example without the need for direct practice or trial and error. Polanyi (1958) further supports this idea, asserting that by observing the master, the apprentice unconsciously picks up the tacit rules of the practice—rules that even the master may not be fully aware of.

These tacit discourses involve strategies that are local, segmentally organized, context specific, and dependent on maximizing encounters with experts and in situ activities. Such knowledge is acquired not through learning procedures of investigation, instruments of observation, or understanding of the theory, but by acquiring a "gaze"—a particular mode or style of recognizing and realizing what counts as reality (Bernstein 2006). Despite the widespread use of observational learning in cultural design and similar practices, Reitan (2006) argues that learning by watching is undervalued in learning theories.

In *Banaras*, watching or observing each other's actions is a common pedagogical tool and is even explicitly indexed in the local dialect, *Bāṇārasī*. Interestingly, in the

Bāṇārasī dialect, this intelligence of the eye is quantified as *Andāz*—an approximate sense of knowing. While visual comprehension of experiential knowledge like weight, force, or temperature can only be a rough estimate, it is sufficient to achieve a sense of knowing. On the other hand, accurate or precise information about actions and sequences of activities can be easily ascertained by observation alone.

Andāz is a form of judgment, but it is not precise. Here, judgment is not a form of decision making in the conventional sense. It is not dependent on logical rules found within rational systems of inquiry, nor is it founded on strict rules of reasoning. Instead, it is likely dependent on the accumulation of experiences from choices made in complex situations. Although judgment is not irrational, because it follows its own form of intuitive logic, learning to make judgments is not about following steps of a technique, following directions dictated by a method or algorithm, or imposing the a priori constraints of a theory (Nelson and Stolterman 2014, 139).

Apprentices, adept at gathering information through observation, may not always need to participate in the performance of a task. Rogoff (1990) reports that in a weaving factory in Guatemala, the method of learning to use the foot loom involves the apprentice simply observing a skilled weaver for several weeks without asking for explanations. Occasionally, the master may ask the apprentice to fetch a spool of thread, but will not permit them to begin weaving until they feel competent after weeks of observation. At that point, the apprentice becomes a skilled weaver merely by watching the demonstrations performed by the master weaver.

Certainly, it would be an exaggeration to assert that watching alone can lead to mastering the complete know-how of every cultural design. Then what aspects of tacit cultural design knowledge can an apprentice absorb by observing the master, especially in the absence of corresponding hand skills?

In the *Bāṇārasī* community, toy designers constantly watch artefacts—from designs to other paraphernalia on television—and they also watch each other. Toy designers watch each other: young apprentices learn by observing their master, and experienced toy designers watch the work of others to upgrade their expertise. Perhaps the greatest influence of the social world is determining which activities are available to children for observation and participation, and who their daily companions are (Rogoff 1990). Therefore, the result of watching varies depending on who is observing whom.

- When apprentices watch their master, they learn to become members of the toy designers' community, following its shared conventions, canons, and beliefs.
- When the masters watch an apprentice, they correct, rectify, or applaud to guide the apprentice in becoming a member of their toy designers' community.
- When practiced masters watch one another, they learn to enhance, upgrade, or amplify knowledge in the toy designers' community of which they are already members.

Marchand (2016, 18) argues that becoming an expert cultural designer is synonymous with being aware of the resources available in the *Karkhanās* and knowing how to orchestrate and competently exploit them when problems arise. Therefore, when the apprentice is given a chance to watch, the offer is not limited to a single

Karkhanā but is deemed legitimate across the entire community of toy designers. The apprentice is granted access to other *Karkhanās*, which furthers the prospects of learning by observing a rich variety of visual stimuli. This learning is not confined to a particular skill, but extends to divergent skills mastered by other community members. This approach introduces a holistic view of artisanship and the knowledge of interdependencies between various skills and skill holders, as evident in the following instances of in situ learning observed in the *Bāṇārasī Karkhanās*.

4.1 Watching Canons, Conventions, and Beliefs

Learning a technology of practice entails not only mastering the use of tools but also acquiring an understanding of the history and culture embedded within the practice. It involves developing the capability to engage more profoundly with the community of toy designers (Lave and Wenger 1991). As they become integrated into this community, apprentices gradually assume active and responsible roles within the social organization, the structure of production, the economic system, and the ritual and ideological framework. These elements collectively and implicitly shape the conventions, canons, and beliefs that underpin the toy designers' community, around which the knowledge of practice is built.

The canons of representation, conventions of production, and the beliefs held by the community concerning a toy are forms of tacit knowledge residing in both the hands and minds of its members. Norman (2013) identifies these as the visceral, behavior, and reflective levels of a design object, respectively. Within the *Karkhanā*, the masters rarely articulated these elements verbally; instead, they remained continuously visible in the making and use of the *Khilōnās*. Often, mere observation was deemed sufficient for the transmission of this knowledge.

In this context, the practice of toy design transcends the mechanical aspects of production; it is a deeply rooted cultural activity. Knowledge transmission occurs through immersive experiences, where apprentices are expected to absorb and internalize these canons, conventions, and beliefs by closely observing the intricate details of their practice. The subtle gestures, the rhythm of work, and the nuanced decisions made throughout the creation process are all integral to this rich tapestry of learning, where unspoken knowledge is as vital as visible techniques.

4.2 Watching Contexts of Representation

In the vibrant world of *Bāṇārasī* toy making, identity and meaning are woven through distinctive shapes, forms, and colours—designed either for ornamental purposes or as expressions of a deeper belief system. Within the community of toy designers exists a shared knowledge of these canons—an unspoken understanding of motifs, shapes, and patterns that recur in their creations. This visual lexicon is ingrained in every toy designer through shared experiences, historical moments, and cultural contexts.

To observe a master at work is to gain an entry point into this world of representation. The ease with which the master navigates these canons of design reveals an implicit knowledge, one acquired through years of experience. Yet, to truly master the skillset requires more than mere observation; it demands that the apprentices emulate and practice, honing their skills through the rhythms of repetition and refinement.

4.3 Watching Methods of Production

The evolution of a *Bāṇārasī Khilōnā* is a collaborative journey shaped by the cumulative skills of successive toy designers who build on the innovations of their predecessors. By closely watching the master, one can discern the sequential choreography of tasks and subtasks involved in designing and manufacturing a toy. An apprentice, through keen observation, learns to gauge the intricacies of tracing patterns, cutting samples, manipulating tools, and applying colours—all integral components of the design skills required to master the skill set.

Unlike formal educational systems, where learning often follows a structured and sequential curriculum, the learning process in this context is more fluid and improvisatory. Apprentices do not receive explicit instructions or feedback; instead, they learn by observing the master's actions as they create artefacts as part of their livelihood.

In the absence of a standardized curriculum, apprentices are routinely exposed to a spectrum of skill levels as they navigate their daily routines within the *Karkhanās*. The master's skills and their application are not always witnessed in the sequential order of their occurrence during the processes of design and manufacturing. Instead, they are encountered as isolated, disparate phases within a lengthy causal sequence composed of interconnected but smaller episodic segments. This fragmented exposure places an additional responsibility on the apprentice: to observe these disjointed sequences across varying contexts and times, to recall and assimilate this nonsequential information into a coherent and meaningful sequence, and subsequently to apply it appropriately to the problem at hand. Observation, in this setting, becomes an inherited tool of learning—one that enables the identification, filtration, mental recording, and retrieval of information during relevant situations while also organizing categories of reality and structuring methods for their application (Rogoff 1994, 70).

Some tasks are subtle and require precise attention to detail. For instance, determining the distance between two holes that dictates the oscillation of a *Dāna chiṛiyā*'s head and tail can only be fully comprehended through practice. However, by watching, apprentices gain a heuristic grasp of the measurements, iterative refinements, and acceptable thresholds of functional accuracy.

Certain tasks are immediately apparent, such as the specific ways tools are wielded and their impact on the wood. This tacit knowledge is rooted in community-held conventions that inform decision making and shape the creation of each toy. Notably, the apprenticeship does not follow a rigid curriculum but unfolds organically, guided by the exigencies of client orders and the complexities of each process. Apprentices often encounter varied sequences as demonstrated by different masters within the *Karkhanā*. The discontinuities in observation, sometimes caused by other duties assigned by the master, necessitate that apprentices stitch together disparate pieces of information into a coherent understanding over time.

4.4 Watching Beliefs About Production and Consumption

Cultural design communities are woven together by intricate, socially constructed webs of beliefs that are essential for understanding their practices and products. Rites and rituals are communal activities and publicly accessible to the novice who is becoming part of the community. These include festivals, ceremonies, and daily customs observed within the *Karkhanā*, such as offering prayers to Lord *Viśvakarma* before commencing work.

These beliefs inform both the production and consumption of *Khilōnās* and link the toy designers and their creations to a broader cultural and spiritual narrative. Through participation and observation, apprentices not only learn the technical aspects of toy making but also absorb the values and beliefs that infuse their practice with meaning and purpose. This holistic engagement shapes their journey from novice to master, embedding them within the community's rich generational knowledge of toy making. The transcriptions below are suggestive of the same.

I was born in the family of *Viśvakarma*. Artisanship is within the house, that is how the child learns.

Transcription 4: XYZ *Karkhanā* 5, Khojwa

The *Viśvakarma*s (caste) pray to Lord *Viśvakarma*, but there are many others (other castes) who pray to him too. We are the original cultural designers of the world. *Viśvakarma* is the architect of this world, so we, his progeny, also create objects.

Transcription 5: XYZ *Karkhanā* 6, Kashmiriganj

Within the community of *Bāṇārasī* toy designers, the interplay of beliefs, values, and notions intricately shapes the relationship between toy designers and their creations. These beliefs manifest through social and religious values, taboos, and superstitions and guide both the practice and the actions of community members. Some of these beliefs are drawn from scriptures and texts, while others are conveyed orally or observed in familial and social contexts.

Rituals and rites often involve specific adults or castes, and for those on the periphery, observation becomes the primary mode of internalizing these customs. By watching, one learns who performs each ritual, when and where it takes place, and the participants involved. For instance, a toy designer is conventionally forbidden from applying the same *Tilak* on their forehead as that of a *Brāhmaṇa*. Yet, through careful observation, toy designers acquire the ability to recognize, differentiate, and eventually replicate the *Brāhmaṇa's Tilak* with precision.

The dynamic interplay between observation and representation is crucial; what remains unseen is rarely depicted in the *Khilōnās*. For example, in a bustling town like Varanasi, the presence of an owl is a rare occurrence. Consequently, this seldom-seen bird is infrequently represented in the *Khilōnās* (refer to Transcription 6). Moreover, the owl carries unfavourable connotations within community beliefs, further diminishing its appearance in the repertoire of toy designers.

I have never made an owl because the owl is the vehicle of *Lakṣmī*. No one keeps it in the house. It is considered inauspicious for business.

Transcription 6: XYZ *Karkhanā* 3, Laksha

According to Bandura (1971), when certain behavior are negatively sanctioned, learning may rarely translate into overt performance. This holds true for instances when non-*Bāṇārasī* customers commission an owl toy; in such cases, toy designers must borrow visual motifs from other cultural practices to create the design. This approach is similarly applied to figures like penguins, kangaroos, or butlers—characters that do not originate from the *Bāṇārasī* visual culture or its broader cultural environment. Consequently, these representations do not embody the distinct *Bāṇārasī* identity inherent in the *Khilōnās*.

The trajectory of an apprentice from the periphery to the center of the community of toy designers is significantly influenced by their understanding of such cultural beliefs. For instance, in *Karkhanā* 4, two aspiring apprentices, who had transitioned from fields as diverse as agriculture and rickshaw pulling, found themselves gradually absorbing the canons of representation and the conventions of making. However, they struggled to fully internalize the beliefs practiced by the masters because they lacked direct access to the rituals and practices embraced generationally by the family. This gap in cultural immersion was evident in the quality of their artisanship and the limited diversity of the artefacts they produced. Their work was often described as *chālū kām* by the masters—a term denoting a standard of design that barely meets community expectations and is thus sold cheaply. These apprentices, lacking both the deep-seated knowledge of community beliefs and the intrinsic motivation to achieve excellence for its own sake (Sennett 2008), rarely progress to become central members of the community of toy designers.

Through this lens, the journey of an apprentice is not solely about acquiring technical skills but also about navigating the intricate social and cultural fabric that defines the practice of *Bāṇārasī* toy designing. The pathway from novice to master is as much about understanding and embodying these shared beliefs as it is about honing one's hand skills. This intricate web of beliefs not only influences the tangible aspects of toy design but also reflects broader cultural narratives and embeds each toy design with layers of meaning that resonate with the shared values and generational habits of the community.

4.5 Watching to Kindle an Aspiration for Advanced Design Skills

By observing the masters struggle with the design of a particular toy, or by witnessing the ease with which they complete another, the apprentice gains an understanding of the complexity inherent in the cultural design practice. As the apprentices begin to practice making *Khilōnās* themselves, the master's watchful eye serves as a guide, gently correcting and refining their technique. This process of correction is not merely about achieving technical proficiency; it is about instilling an appreciation for the care and attention that define high-quality artisanship. Should this value not be implicitly absorbed, the master's vigilant gaze will rectify it when the apprentice later attempts to make a toy as noted in the transcript below.

I stop my children when I see them making too many *Local Khilōnās* (low-skilled, low-paying). I want them to do *Fancy kām* (high-skilled, high-paying).

Transcription 7: XYZ *Karkhanā* 3, Laksha

In this process, the apprentices become aware of the benchmarks by which their work will be evaluated in the future, while the masters continuously strive to push these boundaries through their own practice.

4.6 Watching Responsible Participation in the Community

As discussed earlier, the creation of a single *Khilōnā* necessitates the collaboration of multiple experts. Each toy designer must engage in negotiations with others to fulfill a consignment, a process that is fully visible to the apprentices. Through careful observation of the master's interactions—conversing, bargaining over payments, and adjusting practices to accommodate other stakeholders or clients—apprentices gradually develop an implicit understanding of the cultural nuances involved. They quickly grasp the appropriate diction, the relevant questions, and the behavior deemed legitimate or illegitimate in specific activities (Brown et al. 1989).

Beyond the observable power dynamics in these negotiations, apprentices also discern the subtleties of control, the shifting social boundaries, and the terms of negotiation. They learn to navigate the complex web of communication across different relationships: master-to-master, master-to-client, master-to-apprentice, master-to-community, and master-to-kin. As Lave and Wenger (1991) suggest, such observation deepens the apprentice's comprehension of the contexts in which masters collaborate, collude, and collide, either to bolster their own standing or to challenge that of another.

This mode of learning emphasizes the importance of observation as a foundational tool—an essential "technology"—for acquiring the skills necessary to become a competent member of the community. The act of observant watching functions as a form of peripheral learning, consciously or unconsciously practiced by novices who are allowed to absorb and be absorbed in the community of practice (Lave and Wenger 1991). It is also a strategy employed by those who have achieved central membership within the community of toy designers; it is a means of legitimizing their ownership of the community's knowledge, skills, and resources. These learning trajectories are extensive and demand a keen and patient eye from both apprentice and master alike.

The onus of distinguishing relevant information from irrelevant rests entirely with the learner, as does the self-assessment of what has been acquired and what remains to be learned, despite the absence of a formal syllabus. The learners, whether positioned peripherally or centrally, drive their own progression within the cultural design practice, dominating the learning situation.

The ability to observe effectively is not just about watching passively; it involves an active cognitive process where apprentices must structure and apply what they see to new situations. Rogoff (1994) emphasizes that observation in this context is a sophisticated learning tool that enables apprentices to make sense of reality and decide on appropriate actions based on their observations. Thus, the ability to observe, recall, and integrate nonsequential data into meaningful practices is a critical component of the learning process in the *Bāṇārasī* toy-making practice.

Observational learning, as posited by Bandura (1986) and further examined by Gaskins and Paradise (2010), emerges as a dynamic mode of knowledge acquisition. Within the *Bāṇārasī* toy designers' community, apprentices engage in learning devoid of explicit external input or feedback, a process that demands active participation and improvisatory, practice-based teaching. The master engages in the act of creation not as a formal demonstration but as an integral part of his livelihood, which makes teaching an incidental rather than a focal element of the master–apprentice relationship. In this context, learning to observe becomes a crucial tool for the transfer of the necessary knowledge to acquire the content, rather than the content itself. The fundamental prerequisite for participation in this community of practice is not an existing skill set or formal qualification but the ability to master the primary technology of learning itself—observation.

5 Emulation: Practicing Footsteps to Be a Master

While watching provides an approximate comprehension of an activity, it does not suffice for mastering a cultural design practice. Lave and Wenger (1991) argue that learning within a single community of practice taught through hands-on legitimate peripheral participation develops alongside the ability to perform tasks. They suggest that a training program consisting of instructional settings detached from actual performance tends to separate the learner's ability to manage the learning situation from their ability to execute the skill. Consequently, an apprentice may excel at navigating the learning environment but may never actually perform the skill. In contrast, an apprentice relationship, where the learner actively engages in performing to learn, effectively mitigates this issue.

The sequential nature of observational and emulative levels are outlined by Kitsantas et al. (1999) and Kitsantas et al. (2000). They contend that learners who achieve high emulation accuracy, supported by a model, develop a motorically detailed representational standard that guides and self-monitors their practice efforts and enhances their self-efficacy. This heightened self-efficacy enables them to practice more effectively on their own than learners who have only acquired observational skills.

Sennett (2008) similarly elucidates that around ten thousand hours of experience are required to cultivate a master carpenter or musician. As apprentices accumulate hours of practice, their skills become more problem attuned, whereas novices focus primarily on making mechanisms function. This problem-solving skill is rooted in how apprentices and masters learn to practice and learn from practice—reflecting on and in action.

However, achieving reflection on and in action is challenging for novices who struggle to observe the subtle actions of a master or identify the corrective adjustments needed to align symbolic models with overt performance. In everyday learning, individuals often attain a rough approximation of new behaviour patterns through modeling and refine them via self-corrective adjustments based on informative feedback from performance (Bandura 1971). Thus, before engaging in problem solving, apprentices strive to emulate the master accurately. Reflection on and during emulation confirms and solidifies the approximate knowledge gathered through observation into a habit of the hand. Doing by emulation is the process of transforming the explicit experience of the master (expertise of the master accessible for imitation) into the explicit skills of making an artefact by the apprentice (learning of the apprentice accessible for evaluation).

In the *Bāṇārasī* community of toy designers, emulation serves as a vital method for learning, iterating, transferring, and preserving knowledge. In the local *Bāṇārasī* dialect, the act of copying is referred to as *Nakal karnā* or *Nakal uṭāranā*. This generational habit enables toy designers to replicate cultural designs that date back over a century, even in the absence of oral or written records. Community members observe each other's actions, internalizing design through continual repetition and thereby facilitating the tacit transfer of designs and expertise across the entire community. Design evolution relies on repositories of such "genetic codes" (Jones 1970). Consequently, the knowledge that is known within the community is collectively practiced and, therefore, collectively owned among the *Bāṇārasī* toy designers.

We have to copy each other. If one toy designer has learned to make an artefact, by copying him everyone else also learns.

Transcription 8: XYZ *Karkhanā* 2, Khojwa

Apprentices and masters engage in the emulation of prototypes assigned to them from various sources, such as an old sample preserved for reference, a prototype crafted by the master, or a design provided by a client. This practice of emulation primarily addresses the production demands of the market. Toy designers imitate each other's designs to facilitate the transfer of know-how from master to apprentice for learning purposes, or among master toy designers for skill enhancement.

Beyond this, they also emulate designs from client orders as a means of livelihood. However, emulating client orders can limit the dissemination of knowledge among toy designers, as clients often protect their designs from competitors. These three types of emulation—from masters to apprentices, between masters, and from client orders to toy designers—highlight the diverse motivations and constraints within the cultural design practice.

5.1 Emulation of Masters by Apprentices

In the *Bāṇārasī* practice of toy designing, the pedagogical process known as doing by emulation serves as a cornerstone in transforming observational learning into the apprentice's concrete skill set. Within this dynamic, emulation functions as both a method of acquisition and a means of transmitting knowledge. This dual nature of emulation allows the apprentice to mirror the master's actions, even in the absence of explicit verbal instruction, and transforms these observations into practical skills. Through this mimetic engagement, apprentices internalise the master's techniques, swiftly recognizing and rectifying errors through subtle cues such as disapproval or correction.

From a situated learning perspective, this process is not merely about skill acquisition but also about integrating the apprentice into the community of toy designers, thereby gaining membership and identity through shared generational knowledge (Lave and Wenger 1991). The production of *Khilōnās* within the *Bāṇārasī Karkhanā* reflects this deeply rooted generational practice because community-held conventions and beliefs are meticulously preserved, particularly in *Khilōnās* created for religious purposes (*Mūrti*), for play (*Khelne kā Khilōnā*), and for use by modern customers (*Decoration kā Khilōnā*) (Patil 2014). The continuity of *Bāṇārasīpan*, or *Bāṇārasī*-ness, is maintained across generations, despite shifts in technology and consumer preferences, precisely through this emulative pedagogy.

For instance, the enduring practice of depicting *Kamalanayanī*, or lotus-like eyes, on all figures—both human and animal—has been a hallmark of *Bāṇārasī Khilōnās* since the early 1900s. Although the community may not provide a rationale for this convention, it persists through emulation. Yet this should not be mistaken for mere replication without thought; emulation inherently allows for innovation and adaptation from the original to the replica. Consequently, while the *Kamalanayanī* eyes are traditionally emulated, there remains room for the evolution and enhancement of this design canon over time (refer to Transcription 9).

Older toy designers did not make eyebrows of the Kamalnāyan. But my teacher made it, and I followed him. It looks beautiful.

Transcription 9: XYZ *Karkhanā* 5, Khojwa

5.2 Emulation Between Masters

A local entrepreneur recounted how the popular practice of transforming small wooden birds into earrings began. This idea was initially suggested by a foreign client and introduced a concept previously unknown to the community. Based on oral instructions, the toy designers taught themselves the craft of designing these earrings—a skill they had not practiced before. They learned to intricately design tiny birds using a fret saw, paint them, and pierce them with thin wire. About eight years ago, only a handful of toy designers were capable of creating these designs. However, today, the entire community is involved in producing a wide array of earrings featuring birds and animals in vibrant colours to satisfy increasing client demands.

In this scenario, other community members learned by emulating the toy designers who initially fulfilled the production order for the local entrepreneur. Within the community, emulation is openly practiced due to the absence of copyright restrictions. It is a widely accepted method of "producing" a toy among toy designers. The toy designer community views production in two distinct ways: first, as a collaborative method where a group of functional specialists assembles a single toy; and second, as a strategy for scaling up production to meet client demands by employing multiple groups of functional specialists (refer to Transcriptions 10 and 11).

Designs are often copied. Copying is permitted because a new toy is learned by only one person. When others copy, everyone has a sample, and that is how everyone learns.

Transcription 10: XYZ *Karkhanā* 6, Laksha

I have designed all these *Khilōnās* on display. This toy (pointing to a nested doll with Gandhi's three monkeys on it) was also made first in our *Karkhanā*. But now it is made by others, too. After all, they (clients) want production. Even if they give the order to me, I will not be able to complete it. A single piece or a small order can be completed easily. However, if you must make a thousand pieces, no one can make it alone.

Transcription 11: XYZ *Karkhanā* 6, Laksha

Certain forms of emulation, such as those necessary for production, require precise imitation without any deviations. However, other forms encourage the master or more skilled apprentices to adapt the design to be emulated. These adaptations may be intentionally implemented to introduce novelty, incorporate new or more cost-effective materials—such as the replacement of natural colours with industrial enamel colours, as shown in Fig. 4—or to address evolving consumer preferences. For example, carved animals traditionally bought to decorate the *Jhāṅkī* (a decorative tableau depicting stories of Lord *Kṛṣṇa*) are now also given to children as playthings.

Some emulated modifications can also arise from mistakes. These modifications are considered errors if they alter the defining features of the design. However, if these changes are continuously emulated in subsequent designs by other members,

Fig. 4 Use of lacquer and enamel paint on wooden animals. *Image source* Author

they may gradually gain acceptance. For instance, the three-petal flower motif, which is replicated on nearly all *Khilōnās*, has over time transformed into three individual circles, thereby altering the original design (refer to Fig. 5). Interestingly, this change is evident not just in a single toy but across all *Bāṇārasī Khilōnās*, suggesting that it has been widely emulated by all the toy designers in the community. This transformation further contributes to the evolving *Bāṇārasī* identity of *Khilōnās* in general.

Fig. 5 Changing three-dot motif over the years. *Image source* Author

On the other hand, not all features of emulation are derived from precedents within the toy-making practice itself. Some emulations retain and transform borrowed features from lateral sources as well. For example, visual motifs from Mehndī designs applied to the feet of *Bāṇārasī* men and women by the Sharma community (a community of barbers, not toy designers) are emulated by the toy designers in their designs (refer to Transcription 12).

The ritual of coloring the feet with *Bhūknī* is performed by women of the Sharma family before any auspicious occasion, such as a marriage or *Pāṭh* (reading of a holy book). The *Payal* (anklet worn on the feet by women), like decoration on the wooden toy elephants, is borrowed from these Mehndī designs.

Transcription 12: XYZ *Karkhanā* 6, Kashmiriganj

Beyond visual features, the community also emulates written and oral narratives from the *Purāṇas* (ancient holy texts) and epics like the *Mahābhārata*, transforming them into three-dimensional wooden *Khilōnās*. This process challenges the toy designer to interpret and transform words into images. Some stories, such as *Pūtanā Vadh* and *Nāg-Nāthiyā*, have been generationally emulated, while others, like *Kṛṣṇa* stealing the clothes of the *Gopis*, have been recently added. The emulation of stories that are already part of the worldview and their subsequent adaptation into 3D *Khilōnās* further perpetuates *Bāṇārasīpan* (refer to Fig. 6).

5.3 *Emulation of Client Designs by Masters*

Besides preserving the *Bāṇārasīpan* of *Khilōnās*, emulation also enables master toy designers to earn their livelihood by producing novel designs, either provided by clients or introduced through developmental interventions by NGOs or the government. Non-*Bāṇārasī* clients appreciate toy designers for the precision with which they can replicate a given sample design, and they are paying primarily for imitation rather than innovation. However, since modern client orders often do not align with the *Bāṇārasī* worldview, toy designers, despite excelling in replicating client designs, consciously refrain from integrating non-*Bāṇārasī* features into their own toy-making practices.

Often, clients commission a two-dimensional (2D) image to be designed into a three-dimensional (3D) wooden form. While the client insists on replicating the 2D image, the toy designer must interpret the flat surface in a three-dimensional context, which requires a nuanced understanding of how to translate the design into a tangible form and which demands significant skill and judgment. Sometimes, details like dimensions, full profile, and mechanisms are specified, but there are also instances where incomplete information is left to the imagination of the toy designer. For example, a client once instructed a toy designer to replicate a paper design shown in Fig. 7 in wood, specifying only its height (Height = 10 cm), and insisting that no change be made to the design. The toy designer was expected to determine the remaining dimensions on his own. Despite the lack of design information, the toy

Fig. 6 Depiction of the *Kāliyā Mardan* story in wood. *Image source* Author

designer successfully created the toy, as seen in Fig. 7, including the English text. (As noted in previous chapters, toy designers often have a low literacy level and do not know the English language.)

Emulating 2D instructions to create 3D *Khilōnās* presents challenges not only in visualization but also in reverse engineering, particularly when dealing with intricate mechanisms. An example of this is the pecking toy, which was introduced to the *Bāṇārasī* market from Eastern Europe in the early 1930s. Significant cognitive effort was required to reconstruct the rotary mechanism that allowed multiple birds to move and peck in synchronization on a single plate. Such mechanical movements are rarely

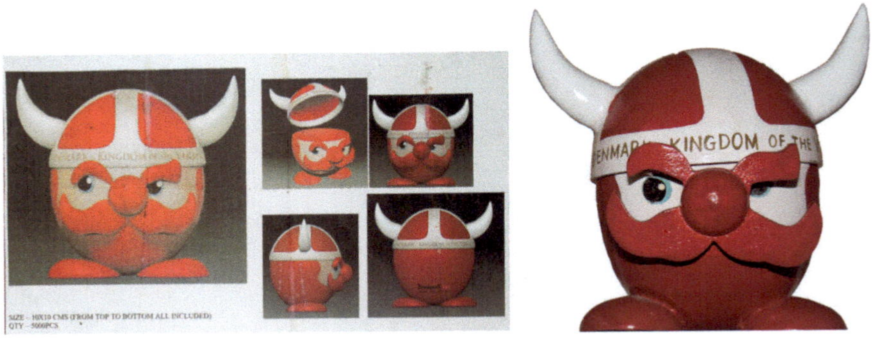

Fig. 7 Emulating a 2D client design into a 3D toy. *Image source* Author

seen in *Bāṇārasī Khilōnās*, which makes its successful emulation by *Bāṇārasī* toy designers all the more remarkable.

The process of reverse engineering the *Dāna chiṛiyā*'s mechanism illustrates the depth of skill and understanding required within the toy designer community. Once the knowledge and skills needed to create the mechanism were acquired, they became freely accessible for other toy designers to emulate. This openness to sharing knowledge facilitated the upgrading of skills and the dissemination of new techniques among all community members.

Upgrading the skills of other toy designers was essential not only to meet the increasing production demands but also because a new toy needs community approval if it is to survive and thrive. In a community that lacks formal records of origin or design trajectories, a feature that is not emulated eventually fades away. The absence of copyright laws within the community allows for the willful emulation of features, which in turn serves as the community's sanction for their continued use.

Once a design is sanctioned through collective emulation, it becomes part of the shared generational knowledge of the community. All toy designers within the community can claim the design as their own—not to mislead the trajectory of its origin, but to acknowledge and celebrate the collective knowledge and skill that produced and preserved it. Thus, emulation in the toy designer community signifies not only the acceptance of a borrowed attribute but also its preservation, improvization, and dissemination.

Emulation plays a vital role in the ongoing evolution of *Bāṇārasī* toy making. It ensures that innovations, whether mechanical or aesthetic, are integrated into the community's repertoire, which enhances both the cultural design practice and the cultural identity it represents.

Numerous studies have drawn a clear distinction between *copying*—often perceived as an unreflective act of imitation—and *emulation*, understood as a discerning and intentional practice (Byrne and Russon 1998; Guth 2010; Nielsen 2006; Tomasello 1990). This chapter brings to the fore the intricacies of emulative practices within the learning processes of a community of toy designers, thereby emphasizing their significance in contemporary pedagogical discussions. The decision making involved in emulating both familiar and novel toy designs within the *Bāṇārasī* tradition reveals that emulation extends beyond mere skill acquisition. It emerges as a cognitive habit refined through thoughtful and deliberate practice, which underscores its role as an advanced mode of learning and creation.

6 Unsituating the Discussion of Situated Learning

Unlike formal educational settings, where knowledge is often explicitly taught, the tacit knowledge embedded in the master's practice is internalized by the apprentice through observation and emulation in the *Bāṇārasī* community of designers. This method allows the apprentice to grasp the cultural design practice not by learning articulated principles but by engaging in the situated artisanship of the master. This chapter argues that in situ observation and emulation create a fertile environment for the co-production of knowledge and emphasizes the importance of these often overlooked modes of learning.

The concept of "authentic" activities, as explored by Brown et al. (1989), posits that knowledge is both situated within a specific context and progressively acquired through practice. This approach necessitates that knowledge be tailored not only to the unique environment of the *Bāṇārasī* toy designers' community but also to the individual learning potential and methods of the apprentice. In *Bāṇāras*, the learning process is not delineated by marked milestones; rather, it is self-paced and aligns with the apprentice's personal learning trajectory. Frequently, the learning curriculum evolves directly from production orders, which contributes to its inherent flexibility and adaptability. Thus, the curriculum is not a preordained structure but is instead shaped by market demands and recognises the apprentice as a toy-designer-in-becoming rather than a toy-designer-to-become.

Furthermore, both masters and apprentices are immersed in the same cultural milieu, albeit to different extents, which grants apprentices access to an unfiltered version of the adult worldview and practice. In the absence of formal examinations or graduation ceremonies, the master relies on continuous evaluation, guided by experience and judgment, to determine when an apprentice is ready to practice independently. For apprentices who are immediate family members, the transition from the periphery to the center of the practice is even more gradual, often lacking any distinct markers that signify the shift from apprentice to master. The apprentice remains engaged in the work, gradually becoming a full participant in the community, and eventually inherits the *Karkhanā*—the very workshop that serves as both alma mater and the foundation of their livelihood.

Within the community's intricate system of knowledge transfer, specific local standards of intelligence and creativity hold significant value. These standards, ranging from the basic requirement of designing generationally practiced *Khilōnās* to the mutual admiration of finely honed carving skills, and even the aspirations of some members to produce automata, are subtly imparted to the apprentice. The community implicitly shares and ingrains these benchmarks, thereby shaping the apprentice's journey. Additionally, the community does not shy away from criticizing subpar work; expressions like *chālū kām* serve as linguistic tools for maintaining quality and introspection within the practice despite their informal and often unspoken nature.

Moreover, these standards extend beyond mere artisanship; they encompass the broader conduct of the practitioner. A *Bāṇārasī* toy designer must not only master the art of toy making but also learn how to responsibly navigate family and social life. Thus, gaining full membership in the toy designers' community is not simply about acquiring a professional skill set; it is a gradual process of enculturation. This process is nurtured by the community's unique pedagogy of learning, whereby the apprentice slowly absorbs the values, practices, and responsibilities that define both the cultural design practice and the life of a *Bāṇārasī* toy designer.

Although this study is situated within a specific Indian community of practice, its insights can enrich the broader discourse on social learning theories, cultural design, and design education. Rather than concentrating solely on pedagogy, formal learning environments might benefit from imparting the skills and tools necessary for learning to learn. Both observation and emulation can scaffold students' ability to learn how to learn and transform them into active learners who pursue their interests while responsibly organizing and managing their learning. In light of the findings from this and similar studies (Bandura 1971; Gaskins and Paradise 2010; Gowlland 2012; Reitan 2014; Rogoff 1990), formal educational pedagogies might consider reevaluating the roles and values of observation and emulation in the classroom.

Looking ahead, it may be necessary to conceptualize a more quantitative approach to studying social learning tools, thereby strengthening the insights gleaned from qualitative studies or theoretical postulations. Most studies on tacit learning remain context specific and qualitative. Capturing tacit knowledge, especially within a contextual living environment, such as the *Karkhanās* of *Bāṇārasī* toy designers, has been an ongoing methodological challenge. The difficulty of communicating and coordinating implicit knowledge in action is widely acknowledged (Gamble 2001; Valkenburg and Dorst 1998). Developing new methodological tools for eliciting and capturing the tacit transfer of information from experts to novices will greatly aid in demystifying the creative processes of problem solving in cultural design.

References

Bandura A (1971) Social learning theory. General Learning Press, New York

Bandura A (1986) Social foundations of thought and action, no 23–28. Englewood Cliffs, NJ, p 2

Bernstein B (2006) Vertical and horizontal discourse: an essay. In: Education and society. Routledge, pp 53–73

Brown JS, Collins A, Duguid P (1989) Situated cognition and the culture of learning. Educ Res 18(1):32–42

Byrne RW, Russon AE (1998) Learning by imitation: a hierarchical approach. Behav Brain Sci 21(5):667–684

Eraut M (2000) Non-formal learning and tacit knowledge in professional work. Br J Educ Psychol 70(1):113–136
Gamble J (2001) Modelling the invisible: the pedagogy of craft apprenticeship. Stud Contin Educ 23(2):185–200
Gamble J (2004) Tacit knowledge in craft pedagogy: a sociological analysis. Doctoral dissertation, University of Cape Town
Gaskins S, Paradise R (2010) Learning through observation in daily life. Anthropol Learn Childhood 85:85–110
Gowlland G (2012) Learning craft skills in China: apprenticeship and social capital in an artisan community of practice. Anthropol Educ Q 43(4):358–371
Guth CME (2010) The multiple modalities of the copy in traditional Japanese crafts. J Mod Craft 3(1):7–18
Howard A (1985) History, myth and Polynesian chieftainship: the case of Rotuman kings. In: Hooper A, Huntsman J (eds) Transformations of Polynesian culture. The Polynesian Society, Auckland, pp 39–77
Jain J (1993) The implicit and the manifest in Indian folk art and mythology. In: Mirror and thread: folk traditions of rural India, pp 46–65
Jones JC (1970) Design methods: seeds of human futures. Wiley, New York
Kitsantas A, Zimmerman BJ, Cleary T (1999) Observation and imitation phases in the development of motoric self-regulation. Unpublished manuscript, Graduate School of the City University of New York
Kitsantas A, Zimmerman BJ, Cleary T (2000) The role of observation and emulation in the development of athletic self-regulation. J Educ Psychol 92(4):811
L'Allier JJ (1997) Frame of reference: NETg's map to the products, their structure and core beliefs. Research and Development—NETg. Retrieved 25 June 2024
Lave J, Wenger E (1991) Situated learning. Legitimate peripheral participation. University of Cambridge Press, Cambridge
Marchand THJ (ed) (2016) Craftwork as problem solving: ethnographic studies of design and making. Ashgate Publishing Ltd.
Muller J (2000) Reclaiming knowledge. Routledge-Falmer, London
Muller J (2012) Reclaiming knowledge: social theory, curriculum and education policy. Routledge
Nelson HG, Stolterman E (2014) The design way: intentional change in an unpredictable world. MIT Press
Nielsen K (2002) The concept of tacit knowledge—a critique. Outlines Crit Pract Stud 4(2):3–17
Nielsen M (2006) Copying actions and copying outcomes: social learning through the second year. Dev Psychol 42(3):555
Nonaka I, Takeuchi H (1995) The knowledge creating company. Oxford University Press, New York
Norman D (2013) The design of everyday things, revised and expanded edn. Basic Books
Patil K (2014) A glimpse inside the Banarasi mind: what do linguistic categories reveal about Banarasi toys? Craft Res 5(2):175–197
Patil K (2015) Designing without a designer: a case study of decision-making in Banarasi toys. Int J Designed Obj 9(3):1
Patil K (2017a) Inexplicit learning: transferring knowledge through visual and emulative practice. Techn Ser Res Sloyd Educ Craft Sci A 24(2):142–164
Patil K (2017b) Dincs as worldviews: things that communicate a mind. In: Toys and communication. Palgrave Macmillan UK, London, pp 137–152
Patil K (2025) Artisanship of risk: the uncertainty of traditional design. J Vocat Educ Train 1–22
Patil K, Athavankar U (2017) Mutable mobiles: charting the cartography of Banarasi toymakers. Des Issues 33(2):43–57
Polanyi M (1958) Personal knowledge: towards a post-critical philosophy. University of Chicago Press, Chicago

Reitan JB (2006) Inuit vernacular design as a community of practice for learning. CoDesign 2(2):71–80
Reitan JB (2014) Learning by watching vernacular Iñupiaq-Inuit design learning as inspiration for design education. Techn Ser A 21(2):1–20
Rogoff B (1990) Apprenticeship in thinking: cognitive development in social context. Oxford University Press, New York
Rogoff B (1994) Developing understanding of the idea of communities of learners. Mind Cult Act 1(4):209–229
Rosch E (1973) Prototype theory. In: Cognitive development and the acquisition of language, pp 111–144
Sennett R (2008) The craftsman. Yale University Press, New Haven
Tomasello M (1990) Cultural transmission in the tool use and communicatory signaling of chimpanzees? In: Language and intelligence in monkeys and apes: comparative developmental perspectives, pp 274–311
Valkenburg R, Dorst K (1998) The reflective practice of design teams. Des Stud 19(3):249–271
Wittgenstein L (1963) Philosophical investigations. Translated by Anscombe GEM. A reprint of the English translation contained in the polyglot edition of 1958. Basil Blackwell
Wolf A (2002) Does education matter?: myths about education and economic growth. Penguin UK

Chapter 6
Artisanship of Risk: The Uncertainty of Cultural Design

1 Artisanship of Risk

David Pye's (1968) exploration of cultural workmanship challenges our perceptions of utility and effectiveness and draws a nuanced distinction between what is "useless" and what is "ineffectual." Much like affection, laughter, or play, Pye argues that craft—although it may be deemed useless—is not ineffectual. To illustrate this point, consider the act of plastering a wall. The utility of plastering lies in its function to prevent the erosion of brickwork. However, the act of smoothing the plaster, though not functionally necessary, is a deliberate, albeit useless, effort. Even more seemingly superfluous is the laying of a fresco mural on wet lime plaster, an act that serves no practical purpose once the plaster has dried and become part of the wall itself. This distinction reflects Louis Sullivan's (1896) dictum that "form follows function," which suggests that anything beyond functional necessity serves no purpose. Yet Pye contends that all workmanship inherently involves the ineffectual; as Richard Sennett (2008, p. 9) aptly states, "the will to do a job well for its own sake."

The worker, therefore, smooths the plaster not to enhance its protective function but to achieve a finish that signifies the quality of the work. Despite using tools such as a hawk board or trowel, the worker may fail to achieve a perfectly smooth surface and risk failure in the process. The task of plastering the wall smooth may seem trivial—a mere expression of craftsmanship—but the designer who conceptualizes this smoothness often has little understanding of the workmanship required to realize it. While the designer's plastering on paper is optimized, the worker's plastering in practice is "satisficing" (Simon 1969). It may not meet the exacting standards of a Computer Aided Design model, but a skilled worker will strive to come as close as possible to the ideal, or *eidos*, of plastering.

This chapter draws upon material from the author's earlier publication "*Artisanship of risk: the uncertainty of traditional design*" in Journal of Vocational Education & Training (2025): 1–22. The current version has been revised for clarity and contextualization.

K. Patil, *Toying with the Mind*, Design Science and Innovation,
https://doi.org/10.1007/978-981-96-9227-9_6

In design education, this disconnect between the maker and the making process is problematic. It is nearly impossible to learn design without engaging in the act of making (Lawson 2006, 6–7). However, when applied, the learning-through-doing pedagogy is often confined to representing, externalizing, or replicating the designer's thought process. This limitation reduces the externalization process, or workmanship, to a predetermined task devoid of decision making or risk. Such constraints are shaped by semantics and limit the scope of cultural design or workmanship as defined by modern standards. From the perspective of tool usage, cultural design is often associated with primitive or low-tech tools or, in more romanticized views, with the absence of tools other than the human hand. Cultural design is also typically seen as the skill of making objects that are decorative rather than functional in contemporary life. The agency of cultural design is frequently attributed to artisans or cultural designers, who are often positioned as lesser than artists and designers. Geographically, cultural design is often located in rural, agrarian societies, rather than urban or industrial ones, which further relegates it to a pre-industrial era. These binaries have likely contributed to the disjunction between hand and mind in modern design practice.

David Pye offered an alternative view. He positioned cultural design (or craft) as a historical and social concept of workmanship, independent of the advancement of tools and machinery or the identity of the maker (Pye 1968, 7–10). He described cultural design as any form of workmanship where the quality of the result is not pre-determined but instead depends on the judgment, dexterity, and care exercized by the maker during the process. Whether performed by a surgeon or a mason, all workmanship shares a common thread: the quality of the result is continuously at risk throughout the making process.

Pye contrasts the *workmanship of risk* with the *workmanship of certainty*, a distinction that is most apparent in the realm of mass production, particularly in the context of full automation. In the *workmanship of certainty*, the quality of the final product is pre-determined with precision well before the actual work is undertaken. There is no gap between the plan and its execution; no fear of deviation or error. A prime example of this type of workmanship is found in the use of computer numerical control (CNC) machines. These machines produce an exact replica of the digital design they are programmed with, which leaves no room for machine error. In this context, a worker who is engaged in the *workmanship of certainty* cannot spoil the job, no matter how complex it may be.

Conversely, a worker engaged in the *workmanship of risk*, even when assisted by an array of jigs and tools, constantly faces the possibility of failure. The distinguishing factor between these two types of workmanship is not the presence or absence of automation or mechanization but rather the degree of risk of failure that the worker is willing to take on. Pye provides an extreme example to illustrate this distinction: consider a dentist drilling a tooth with an electric drill versus a carpenter drilling a piece of wood with a hand-driven drill using a jig. Despite the mechanical nature of the dentist's operation and the manual nature of the carpenter, the dentist assumes a far greater risk. The power source—whether electric or manual—is irrelevant to the level of risk involved. In fact, electric tools sometimes demand more judgment, dexterity, and care than their hand-powered counterparts. Pye therefore poses a simple question to determine the type of workmanship at play: "Is the result predetermined and unalterable once production begins?" (Pye 1968, 7–10).

Pye feared that the *workmanship of risk* would eventually disappear and be supplanted by inferior forms due to a lack of theoretical understanding and established standards. He was concerned that the opportunities inherent in risk would be neglected. Despite Pye's insights, there is still a limited understanding of the role of risk in the learning process, or in the evaluation of the level and type of risk involved. Cultural design is often seen as a mere palimpsest of repetitive actions, with little room for deviation, let alone risk. This chapter, therefore, seeks to explore the nature of risk in cultural design and methods for evaluating it by examining the practice of wooden toy making in *Varanasi*. India can uncertainty in cultural design be characterized as mere imitation of established traditions?

In addition to David Pye's theory of *workmanship of risk* and uncertainty, this chapter also draws on the work of Nonaka and Takeuchi (1995) on the transfer of tacit knowledge. A brief outline of their theory is provided below.

In 1985, the Matsushita Electric Company, based in *Osaka*, faced challenges in designing a new home bread-making machine. Their initial designs failed to knead dough adequately, which resulted in bread that was overcooked on the outside and undercooked on the inside. Despite extensive analysis, including comparing X-rays of dough kneaded by machine and that by professional bakers, the company struggled to find a solution.

Ikuko Tanaka, a software developer, proposed a novel approach. She studied the kneading technique of a baker at The Osaka International Hotel, which was renowned for producing the city's finest bread. During her apprenticeship with the baker, Tanaka observed his unique method of stretching the dough. After a year of trial and error, and close collaboration with the project engineers, Tanaka developed product specifications, including the addition of special ribs within the machine, to replicate the baker's stretching technique. This innovation led to Matsushita's unique "twist dough" method and resulted in a product that set sales records for new kitchen appliances in its first year.

Tanaka's innovation serves as a vivid example of the transition between two fundamentally different types of knowledge. The first type, *tacit* knowledge, is embodied by the chief baker at the *Osaka International Hotel*—the very source of Tanaka's breakthrough. *Tacit* knowledge is notoriously difficult to articulate; it is deeply personal, often resistant to formalization, and challenging to communicate to others. A significant portion of this knowledge consists of the *"know-how"* that a cultural master designer acquires *"at his fingertips"* after years of meticulous practice. Despite possessing profound expertise, such a master often struggles to articulate the scientific or technical principles that underpin his skill. Furthermore, *tacit* knowledge encompasses mental models, beliefs, and perspectives that are so deeply ingrained in an individual's experience that they are taken for granted, which makes them even more elusive and difficult to express.

The challenge for Tanaka and her team was to transform this *tacit* knowledge into something explicit and sufficiently detailed to define the specifications of the bread-making machine. *Explicit* knowledge, in contrast, is impersonal, formal, and systematic, and can be easily conveyed through language, scientific formulas, or coding platforms.

As explored in Chap. 5, Nonaka and Takeuchi (1995) identify four fundamental patterns of knowledge creation based on the interaction between *tacit* and *explicit* knowledge, each leading to a distinct process of knowledge exchange (see Fig. 1 in Chap. 5). The first pattern, *tacit-to-tacit* knowledge exchange, occurs through the process of *socialization*. This involves sharing experiences and thereby creating shared *tacit* knowledge, such as common mental models and technical skills. The second pattern, *externalization*, transforms *tacit* knowledge into *explicit* concepts and allows for more systematic communication. Third, the *combination* of *explicit* knowledge involves synthesizing different bodies of *explicit* information into a coherent knowledge system. Finally, *internalization* occurs when *explicit* knowledge is absorbed and embodied, whereby it is transformed back into *tacit* knowledge within the individual.

Of course, much remains to be understood about the conversion of *tacit* knowledge into *explicit* knowledge within the knowledge spiral. What are the pedagogical mechanisms underlying these *tacit-to-explicit* processes, and how do these pedagogies shape the workmanship of risk and certainty? The learning pedagogy of skill-based practices, such as cultural design, is often hastily subsumed under the broad term *learning by doing*. Yet, as discussed in Chap. 5, cultural designers employ a variety of pedagogical methods. While not discounting the significance of *learning by doing* in the pedagogy of the hand, this chapter will also explore other forms of engagement, including emulation, observation, and reflection in and on action. Despite their contested status, copying, imitating, and emulating play crucial roles in many cultural processes of knowledge transfer. Observation, though frequently mentioned, suffers from the absence of a comprehensive generative theory. Schön (1922) elaborates extensively on *reflection on* and *reflection in action* to describe the work of a practitioner in action; however, the risks inherent in reflection and the criteria for evaluating such risks fall outside the scope of his work.

According to Lave (2011), the acquisition of hand skills occurs iteratively, in cycles of approximation. In each iteration, the learner completes the entire process from start to finish, gradually approaching the level of quality achieved by the master. Each cycle of approximation involves a risk—a gamble on the part of the novice—aimed at emulating the ideal object designed by the master. Similarly, the master in cultural design engages in analogous risks, not only when replicating the work of other masters but also when attempting to materialize an unfamiliar concept conveyed on paper or verbally by a client without the guidance of a tangible prototype. This iterative process of approximation underscores the centrality of risk in both the novice's and the master's journey toward mastery. According to Lave, this practice might be called "successive approximation," in which the whole process of making an artefact is repeated from start to finish until all parts are equally well-made.

> Practicing subassemblies might have been way to reduce mistakes. But practice by successive approximation was a better way to integrate rudimentary skills and a conceptual model into a smooth process of garment production. It could be learned by observing masters and

> advanced learners at work, and it would serve the learners well as they become mature practitioners. Successive approximation was an efficient form of practice so long as the costs of making mistakes were not much higher than they would be if difficult aspects of construction processes were practiced in isolation first. After all, the exceptional skill of these tailors did not lie in their ability to make trousers, but in their ability to make a dozen pair of trousers from start to finish in one day. (Lave 2011, 76)

In each cycle of approximation, certain skills are mastered more effectively than others, which leads to a nonlinear progression in the learning process. This nonadditive nature of skill acquisition necessitates the simultaneous learning of all the skills required to produce an *artefact*. Furthermore, the value of the output is not measured by its ability to seamlessly replicate the desired object but by its approximate closeness to it. In contrast, the value of automated skills lies in their accuracy and precision in replication, which aligns with the *workmanship of certainty*. However, for the cultural designer, the value of the output is determined by its approximate fidelity to the client's order, thereby increasing the inherent risk of *workmanship.*

This chapter will delve into the pedagogies of the hand through the lens of *workmanship of risk* and *certainty,* and will propose a model for the transfer of *tacit-to-explicit* and *explicit-to-tacit* knowledge. My postulations are informed by six years of ethnographic research on the master–learner relationship among *Bāṇārasī toy designers* and my sixteen years of formal pedagogical experience teaching undergraduate and postgraduate design students. I envision this chapter, in conjunction with the preceding ones, as an assemblage of conceptual tools with which to investigate the learning trajectories of both learners and masters within the social structure of their community.

1.1 The Ways In

During my initial visits to the *Karkhanās* of the *Bāṇārasī toy designers*, I found myself puzzled by the apparent absence of a uniform teaching practice—no standardized curriculum existed to guide the schooling of learners across the *Karkhanās*, nor was there a formal system in place for assessing the knowledge imparted to them. There were multiple unstructured entry points into the learning system, and explicit communication between the master and the learners seemed conspicuously lacking. However, after a few weeks, a pattern began to emerge from this seemingly disorganized structure. The pedagogies of observation, emulation, doing, and reflection, as discussed in Chap. 5, revealed themselves as integral to the tacit exchange of knowledge and skill transfer between the learner and the master, and even among the masters themselves within the community. Unlike the explicit tools of knowledge transfer employed in formal schools, which are predominantly teacher centric, these pedagogical tools shift the responsibility of learning onto the student, thereby creating a learner-centric pedagogy. Intriguingly, unlike formal schooling, which typically ends with the award of a degree, learning within the *Bāṇārasī toy designer* community is a lifelong activity. While the process of learning remains consistent, the content evolves as the learner progresses from a peripheral to a core membership within the community. For a more detailed exploration of the learning trajectory uncovered during this ethnographic study, please refer to Chap. 5.

Masters in the *Bāṇārasī* community of *toy designers* provided apprentices with legitimate access to the workplace and the opportunity to move from the periphery to the core of community practice. The master not only determined the way in for the learner but also evaluated and judged the learner's trajectory to determine the way out. Although the explicit teaching of concepts was never a practice followed by the masters, they kept an inadvertent eye on the mistakes and challenges encountered by the learners and corrected them from time to time. Some errors might go unnoticed by customers, but the masters seemed acutely aware of various procedural missteps that could lead to the same problem in a toy. Depending on the extent to which an error impacted the toy's marketability, the master would sometimes correct it or choose to overlook it. Correcting an error also served as a way to communicate the community-held standards of *workmanship* (Lave 2011).

Not all toys emulated by the community members were novel or unfamiliar. In fact, cultural design has often been criticized for merely imitating the evolved forms of the past. Emulating traditions typically requires a shorter learning trajectory than problem solving for novel situations, though the former, as we will soon see, also carries its own risks of failure. The extent and nature of risk, then, is shaped by the structure of the problem, the aspired outcome or value, and the available problem-solving tools, techniques, and strategies, which vary across cultures. The cultural understanding of a problem and the evaluation of its solution shape the nature of the risk involved. The forthcoming sections discuss the acts considered errors, those that can pass unnoticed under an uncritical eye, and the outcomes that are aspired to in the *Bāṇārasī* community of *toy designers*.

2 Is It Possible Not to See What Is Visible?

Superficially, merely observing another person should not be a challenge. Yet, as we shall explore, this ostensibly instinctive skill of observation entails risks of failure. Observation legitimises the learner's participation and exposes the community's practice to the learner's scrutiny. This act of observation is not a passive, vague demand to watch the master continuously. Instead, it is an active and focused process, driven by the anticipation of, or the process of becoming, a central member of a professional community.

Bandura (1986) outlines four preconditions necessary for acquiring new skills through observation:

1. **Attention**: The learner must direct visual, auditory, and sometimes even olfactory attention toward the skill to be learned. In the case of toy designers, this involves close attention to the kinaesthetic memory associated with their craft.

2. **Retention**: The learner must be able to store the know-how in their long-term memory, which ensures that they can recall it when needed.
3. **Reproduction**: The learner must possess the ability to retrieve and physically reproduce the skill or task observed.
4. **Motivation**: The learner must be motivated to persistently practice the skill until it is mastered.

These preconditions also require a master or knowledgeable other from whom the learner can observe and learn. Throughout the apprenticeship process, as noted by Lave (2011), learners in the Vai and Gola communities were peripheral participants with legitimate access to the arena of professional practice. In this context, master tailors and learners in the shops exhibited varying levels of skill, which allowed learners to predict future movements and understand the steps required over time. Learners had the opportunity to observe both the construction process and the finished garments; these experiences provided them with tangible examples produced by others with differing levels of conceptual and physical skill and experience.

An attentive observer can capture numerous nuances of the task at hand simply by watching the master at work. Information regarding postural cues, the sequence of subtasks, the time and effort required for each task, the appropriate tools and machinery for each operation, the ways of operating and holding a tool, changes in attention and gaze direction, the negotiation of task outputs, and the mapping of meaning over form can all be absorbed by the learner. Observational learning is especially applicable to understanding these complex relationships.

Most importantly, by observing the entire sequence of tasks required to achieve a particular end, the learner forms an initial approximate conceptual model of how a specific object is made. This mental model is incomplete and will be filled in when the learner physically engages in making the object. In this first approximation, the learner gains rough insights into the principles of how machines, tools, materials, and design interact with the human hand to achieve a desired result.

Through further emulation, practice, and reflection, these initial observation-based models are refined and strengthened throughout the learning trajectory. The effectiveness of this observation-based model plays a crucial role in determining not only how quickly the learner masters a task but also how proficient a master they later become. What is not observed cannot be emulated, practiced, or contemplated.

For example, most toy designers in *Bāṇāras* are unable to read a sketch, especially one depicting a foreign cartoon-like image. On one occasion, a toy designer was so unfamiliar with a customer's sketch that he mistakenly held the photograph upside down and mistook the inverted image for the correct design. He scrutinized the purported eyes and nose of the figure for nearly six minutes while holding the photograph incorrectly. This incident highlights a critical limitation: what the toy designers cannot accurately observe, they cannot create. This limitation restricts the variety of toys the toy designer can produce.

No matter how attentive a learner might be, there are inherent limitations to what can be observed. The eye, despite its precision, cannot capture everything—especially the fine motor skills that are crucial in artisanship. These fine motor skills often demand a very close spatial proximity to the master, something that is difficult to achieve in the cramped spaces of small *Karkhanās* (workshops). The physical layout and the master's own need to meet production deadlines often prevent learners from getting close enough to observe these intricate details.

As a result, learners are frequently left to speculate intelligently about these fine motor skills. However, this speculation is fraught with risks. For example, abstract physical values such as the exact force applied by the master are nearly impossible to gauge from a distance. Misjudging these can lead to incorrect conceptual models or failures in executing the task. This means that even if the learner is highly attentive, their understanding might still be incomplete or incorrect, which can lead to a higher risk of failure.

The risks extend beyond the technical aspects of learning to the social fabric of the community. There is a significant risk of not performing socially appropriate tasks within the community. In one example, I observed two aspiring learners in Karkhanā 4 who did not culturally belong to the toy-making community. Unlike those who grew up in toy-making families, these learners had shifted from agriculture and rickshaw pulling into toy making. Because they did not live with the masters, they lacked legitimate access to the beliefs, customs, and rituals practiced by the toy designers within their close-knit homes or castes, such as the Viśvakarmas or Prajāpatis.

Understanding and adhering to the socially acceptable values in toy making is crucial for gaining full membership in the toy designer's community. The lack of this cultural knowledge significantly slowed the trajectory of the aforementioned two learners from the periphery to the centre of the community. What they could not observe, they could not emulate, and this failure to assimilate the customs and conventions of toy making hindered their progress.

In this context, the consequences of not learning are subtle yet profound. Unlike formal education systems, where a failing grade might formally terminate a student's progress, in the toy designer's community, poor learners are gradually disengaged. They are not assigned additional responsibilities or work, which can lead to their gradual loss of interest. Eventually, the learner may leave the community voluntarily, without any formal evaluation or termination process. This self-termination is a reflection of the community's organic and unspoken evaluation system—where the learner's decision to leave is as much an acknowledgment of their lack of progress as it is a subtle suggestion from the community that they do not belong.

This organic process of disengagement highlights the delicate balance between technical skill and social integration in cultural design skills. The risks involved in learning go beyond mastering the physical tasks; they encompass the entire cultural and social context in which the cultural design is practiced. Failure to navigate both these aspects effectively can result in a learner being quietly, yet decisively, eased out of the learning community.

3 Is It Difficult to Copy?

Learning by emulation is often used in cultural design communities to replicate the skills of the master or to achieve the design proposed by the client. This pedagogical tool is often criticized for limiting handwork to workmanship of certainty—a predetermined skill lacking any decision making or risk in its operation (Pye 1968, 20). It is perceived as a mindless, mechanical activity that requires at most some dexterity and deftness of the hand to yield the certitude of success. The very definition of emulation indicates the difficulty in failing to imitate a task by any reasonable learner or master.

Primarily, it can be argued that emulation shifts focus from knowledge transfer to knowledge acquisition and thereby puts the onus of learning on the learner instead of the teacher. Such a shift demands multiple learning skills from the learners, who may fail in some of these skills. At the least, the learner must have the initiative, passion, and perseverance to continue on the learning trajectory independent of the lack of motivation provided by the master. The pedagogy of emulation of emulation is initiated and led by the learner, with little or no push from the master. Not everyone may have the drive to continue learning on their own. Moreover, the task of the learner is not only to pursue the movements of the master, but also to understand, memorise, and appropriately recall the sequence of events in different contexts. Conceptualizing the sequence of events will also create a mental model of the activity for the learner. There is a risk in creating the correct mental model through observation or emulation alone because in both learning methods, there is no explicit scaffolding of instructions. Both observation and emulation have a gestaltian tendency to make sense of incomplete information, and, similar to gestalt, it is not the amount of information but the kind of information that will shape the correct mental model. This implies that what, and not how many kinds of, activities a learner decides to observe and emulate, drives the learning trajectory. After all, the whole mental model must be created from sequential tasks of gross motor skills alone. Fine motor skills are difficult to observe and emulate. The effect of one task upon the other also must be mentally computed and executed at appropriate moments in contexts of execution that differ from toy to toy. For instance, threads tied on the tail and neck of a pecking bird in the pecking toy are pulled by the wooden weight rotating below the plate. Although the sequence of activities as well as its effect is understood easily by watching the master, there are several difficult spots in conceptualizing and executing this task. It is difficult to estimate the length of the thread tied to the tail and head in proportion to the length left below the plate. How long should it be? If there is a mathematical formula for the same, then it is not taught. If there is a rule of thumb, then it is not shared by the master with the learner. It simply has to be learned while emulating—and through mistakes.

Moreover, as discussed earlier, it is not enough to learn the length of the thread in this toy. The same mechanism is applied in *Hilāntā* as well as the pecking *flying toys*. How should it be applied to other toys of different shapes and movement? Cross-contextual application of the skill in different categories of toys requires a higher level of mental computation and hand dexterity than emulating a task for toys within the same category. The risk of failure may increase manifold if the object of emulation is unfamiliar. In the case of *Bāṇārasī* toy designers, almost all client orders fall in this category. In most client orders, a 2D paper design is given. Though no design changes are required, the cultural designer needs to emulate the 2D drawing on a 3D surface, which requires substantial skill and judgment as seen in Fig. 1. Not only must the toy designer comprehend the transformation of the 2D design into relief, but he must also imagine and complete the back profile of the toy.

In a learner-centric process, the curriculum is not laid down by the master; therefore, the learner is required to decide what must be learned and how it must be learned according to his disposition. The tasks carried out by the master cannot be emulated successfully in one cycle of approximation. There is a conceptual risk of either not memorizing or not recalling the appropriate sequence of the tasks. Even when a sequence is known, it is not necessary for the learner to know the cause and effect of certain actions, which may lead to errors. The learner must decompose the task into learnable learning chunks compatible with his own ability to learn, approximate, and

Fig. 1 Transformation of 2D design into a 3D surface. *Image source* Author

improvise. This, of course, further implies that a learner may be able to conceptualize correctly but may continue to require several cycles of approximation before he arrives at an acceptable prototype. Therefore, conceptualization of a task may not be realized in its execution until several rounds of approximation have passed.

Deftness of the hand will only improve in each cycle of approximation, but lack of it is a major reason for frequent failure. The two aspiring apprentices from Karkhanā 4 mentioned earlier did not have continuous access to many rituals and practices followed in the core circle of toy designers. This lack of opportunity was visible in the quality of their workmanship as well as in the limited diversity of the artefacts designed by them. Their work was often referred as *challu kaam* by the masters. *Challu kaam* describes a standard of crafting that is barely acceptable in the community and whose crafted items are therefore sold cheaply. These learners lacked the hand skills to emulate correctly. If at all they succeed to emulate correctly, they may have required many more cycles of approximation than other learners in the Karkhanā.

Most importantly, the learner should have the insight to identify and self-correct mistakes while emulating. In the absence of explicit instruction and guidance, the learner must learn to identify errors, find solutions to correct them, and evaluate the proficiency of the devised solution, too. Hence, the risk is not just of committing an error, but the risk also involves the inability of the learner to self-correct the error. Normally, in a formal educational context, this responsibility belongs to the teacher. However, in a learner-centric pedagogy, it is the learners who commit, confirm, and then also correct their own mistakes.

4 Can You Act Without Thinking?

The second conceptual model attained through emulation fills several gaps in the mental model that was left behind while learning through observation. This approximation is much closer to the master's actions and goals because the process of making gives haptic feedback from material and tools, shows tangible errors that either magnify in consecutive steps or lead to a complete failure, provides a kinaesthetic sense of bodily postures and grasps, and provides visible consequences of actions. Learning by doing is a successive cycle of approximation that further refines and scaffolds the learning trajectory of the learner. Unlike emulation, while *doing* the learners are not attempting to emulate, but, instead, are making their own mental model of the process through their actions. In this cycle of approximation, the learners may even deviate from the master to suit their own style or pace of making. However, the learner is yet to master the process. This third cycle of approximation is not additive, but simultaneous learning of skills. No single skill is mastered completely before engaging with another. Among the repertoire of skills, the learner may perform better in some, which can result in a toy that is more finished in some respects than others. Learning, here, is a simultaneous, holistic activity and not a consecutive or linear process.

It is often believed that participants in such learning communities do not possess a creative vigour or any exceptional skill. They create beautiful products by merely adhering to a long-standing tradition and implementing minor improvements only in recognition of a failure. But they are unsuccessful when faced with the unfamiliar and complex task of actually inventing forms from scratch (Alexander 1964). However, it is presumptuous to believe that all learning communities of cultural design lack the ability to innovate. The *Bāṇārasī* toy designers, for instance, show a diversity of form and the ability to innovate for the sake of improvization alone, as well as for meeting client demands (Patil and Athavankar 2022). Similarly, the Honeybee Network founded by Prof. Anil Gupta, Indian Institute of Management Ahmedabad has documented disruptive innovations made by several grassroots designers, such as Mitticool by Mansukhbhai, or the stencil cutting device for hand embroidery designs by Shaikh Mohammad Nazim. Evidently, given the low-tech, low-literate, and less-user-informed processes adopted by such design communities, their path to a well-finished product is strewn with risks.

There are two kinds of approximations possible in the learning-by-doing method. One approximation is a consequence of repeated emulation, in which the learner attempts to make a product independent of the master. Another type of approximation is a product that learners, often advanced learners, or masters attempt to innovate or design on their own or based on client designs. In the former case, the learner may have a correct conceptual model but may not possess the hand skills to execute the process, while in the latter case, the learner may possess neither. The purpose of these approximations is to improve both quality and production speed. For instance, in the client design discussed in Chap. 5 (Fig. 6), the toy designer lacks complete understanding of the design and is unable to replicate it according to client specifications. Partly, the learner has taken the risk of making a design with incomplete information. In this case, the client has not given dimensions of all parts of the toy, which the toy designer is expected to visualise and compute on his own. Also, in the absence of any jigs, the toy designer is taking the risk of painting a symmetrical design of the face, writing a script in English without knowing the English language, and attempting to bring the precision of a CAD design to hand-made objects.

The uncertainty of repeating (that is, emulating oneself or another) is more predictable compared to making something anew or seeking a solution to a problem. During recurrent field visits at Varanasi, none of the cultural designers were seen sketching a solution to a problem. When given a problem, they "made" and not "drew" a "solution." Ratanlal Prajāpati, a toy designer, made an innovative mechanism for moving a traditionally static toy. Instead of a solid carved body, he assembled flat cut-outs to create volume. He modified and assimilated the well-rehearsed two-part mechanism of the *Hilāntā* into a four-part motion, without any drawings or for planning (refer to Fig. 2), guided by habit of action alone. When I critically pointed out the lack of space for grasping/holding the new design, he did not go back to the drawing board to rethink. Instead, he tried to attach an unused wooden cylinder as a pedestal. When one size did not fit, he iterated and fit another cylinder. The modified solution was complete in barely 10 min.

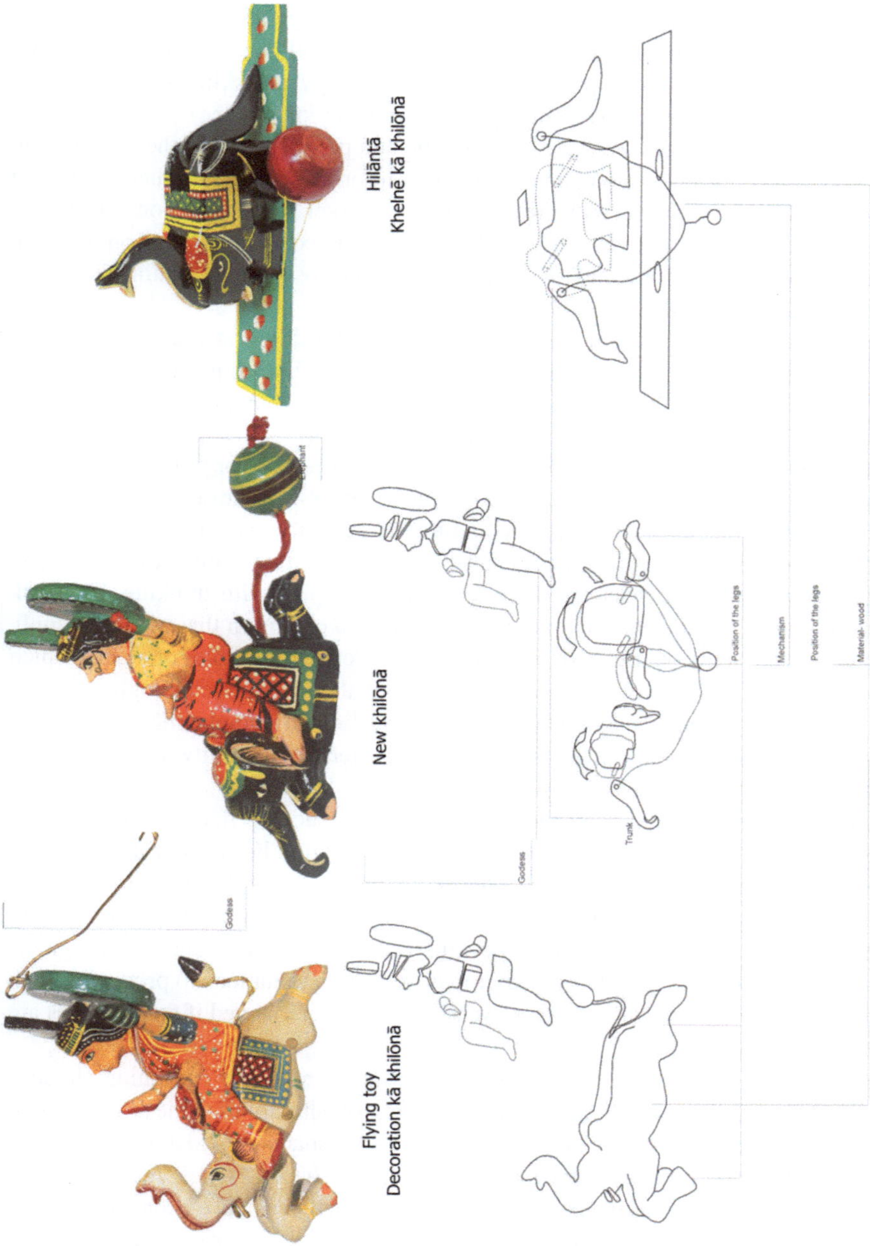

Fig. 2 Assimilation of the mechanism of the *Hilāntā* toy with the form, construction, and meaning of the *Flying toy*. *Image source* Author; also, reproduced in Patil and Athavankar (2017)

In the absence of the skill of designing on paper aided by user surveys, market benchmarking, and other such sources of insights, Ratanlal's thinking runs simultaneous to the hand. There is no time lag between designing and production, but instead the process is a reflective practice in which design and production overlap. Such thinking-enabled production (which assists in worldview acquisition) occurs through the knowledge of available material, technology, and skill wherein thinking moves from the known (possibilities of the available knowledge) to the unknown (a solution that emerges from the combination and permutations of the known). On the other hand, designing on paper is often an informed conjecture that is directed toward an optimal solution with the aid of user, a market, and production data; therefore design thinking often moves from the unknown (a conjecture based on what the available knowledge could be) toward the known (the possibilities within constraints that are made available with time).

The risk involved in simultaneously thinking and doing is much higher than in processes in which the latter follows the former. Sketching and planning, a quintessential stage of the design process, is redundant in *Bāṇārasī* practice because all decisions are made during production itself.

The opportunity of reflecting on design decisions on paper or prior to production saves precious time and material cost for formal designers. On the other hand, iterative prototyping of the artefact allows the cultural designer to make, modify, and refine the design in practice rather than on paper until all constraints are optimized. Here, production is not mindless repetitive labor; it is coupled with thinking. This risk makes change spontaneous and not premeditated—a condition that bears fruit only through habit of action. Iterative prototyping is effective for smaller products, which involve lesser numbers of materials and therefore also slower technologies. Slower technologies (Hallnäs and Redström 2001) have limited complexities that can be handled through the skill of a single person and offer the flexibility of changing the pace of production as and when required.

This risk is further heightened in a community such as that of *Bāṇārasī* toy designers, where skill lies in making a part but innovation lies in making the whole toy. A toy is mostly made from part to whole, which is true for all *pattern*, lathe-turned, or carved artifacts. The toy is assembled by a particular toy designer, but each part is made, and at times improvised or changed, by many individual toy designers. Parts of toys are shared with each other in different combinations to produce design variations. Hands, feet, and at times even faces are interchanged if the figurines are similarly proportioned.

Cultural designers employ this technique of combining and permuting already-existing elements to make both well-known and novel alterations to the artefact. With just minor modifications, the new toy design in Fig. 3 shares the existing mechanism of the *Hilāntā* parrot, its dimensions, as well as its shape—wings, feet, base, and square weight. However, the peacock's neck, head, and beak have been altered to fit the characteristics of the new birds. For instance, the parrot's beak has been elongated and rounded for the duck while it has been curved and sharpened for the crow. Here, combinations and permutations of already-existing components coupled with the incorporation of new elements have led to the creation of new artefacts that are aided by visual painted signs for things such as body colour, eye-related motifs, feathers, and claws. Even if the artefact was put together by a certain artisan, each component was created—and occasionally improved or altered—by numerous different cultural designers.

The innovation here is a result of a cumulative, collaborative, and collective process that is shared and improvised by multiple cultural designers (Leadbeater 2009; Reitan 2006). But unlike the communities of innovation discussed by Leadbeater, the *Bāṇārasī* toy designers innovate un-self-consciously (Alexander 1964) without a shared plan or leader. Therefore, knowledge of the design and its change is a community property rather than an individual one through an implicit, unwritten, and unplanned consensus among the toy designers; it is not an explicitly planned agreement.

Fig. 3 New variations of the *Hilāntā* toy designed by modifying the parts of the existing design. *Image source* Author

5 Can You Think Mindlessly?

Almost all thinking runs parallel to action or doing (except observation); therefore, uncertainties of the outcome of the fourth approximate cycle are closely aligned with the previous three cycles of approximation: namely, observation, emulation, and doing. In fact, the fourth approximation—reflection—is not a sequential consequence of the previous three cycles but a gradual iterative build up, a porous palimpsest of the three previous approximate mental models.

In the observation phase, the first approximate cycle is a mental or conceptual model built with very limited visual knowledge of the action and sequence of a task that is conducted by advanced learners or a master. Hence, all the risks discussed in the first approximate cycle are a part of the reflection of the learner. Similarly, in the second and third approximate cycle based on emulation and doing, the perils of execution based on a mental model that is being continuously refined and corrected are apparent. Often, the risk of approximation not only emerges from the lack of dexterity in the hand, but also from the difference in the worldview of the community and that of the consumer. During a visit to a Karkhanā, I observed that a learner appeared confused regarding the assembly of the tail of a wooden bird. The master reminded the learner that the client's design lacked a tail and that they were required to follow suit. Nonetheless, the learner attached a tail to the bird (please see Fig. 4). The master smiled and disregarded the defiance because it asserted the community worldview, which forbids the representation of a decapitated body. Both the master and the learner had the correct mental model of the client toy, but they openly defied it in order to create an approximate physical model in consonance with their own worldview. For client orders belonging to other cultures and worldviews, the toy designers are often unable to seek any contextual logic. The risk of an inaccurate mental model of the design is high due to the toy designers' lack of user data and cross-cultural exposure. This situation can result in loss of job orders.

Throughout the learning process, the learners mapped their own actions with that of the master's to identify errors, corrected them by emulating the master, and applied appropriate problem-solving strategies for correction. Learners are encouraged to reflect on their own actions and correct the majority of their errors independently, without coaching. Masters are only presented with mistakes that learners cannot fix on their own; this enables the master to concentrate on more significant and persistent issues in learners' skills. Reflection is a part of doing, so the skill of self-correction is acquired almost immediately (which may explain the random and restricted nature of the master's coaching). All these processes of error recognition and correction are a result of the reflective ability of the learner, which, in the case of hand-based learning runs parallel to doing.

Moreover, during reflection, the learner memorises not the smallest meaningful task, but instead the smallest series of tasks that make a meaningful sequence. This form of retention facilitates combinations and permutations of various parts and processes of making an object to create novel varieties, but it also explains the rigidity toward changes required in creating a novel form or shape or the errors

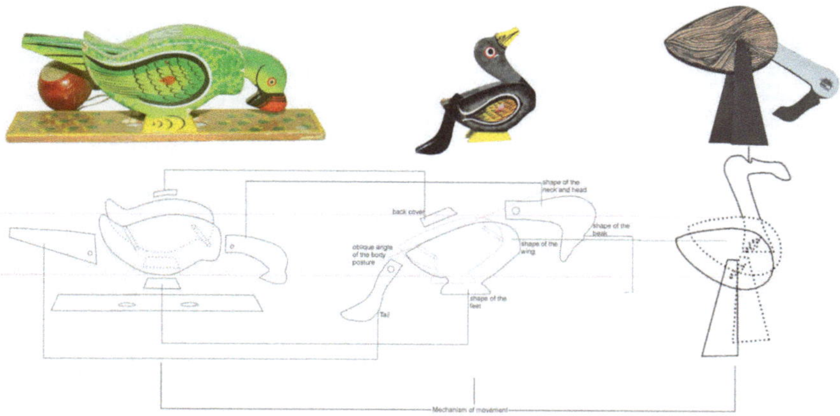

(a) *Hilāntā* parrot (b) New Design (c) Client design

Fig. 4 New design (**b**) created by assimilating the features of *Hilāntā* parrot (**a**) with the Client design (**c**). *Image source* Author

that occur in cross-segment application of knowledge. Jones (1970) considers this rigidity to be borne out of traditions, rituals, and taboos, but it is also an outcome of the learning process of which the smallest unit is a segment of meaningful tasks rather than singular context-independent actions. To create a novel shape, a series of meaningful tasks must be revised or rejected, and a new series of tasks must be learned. Hence, the resistance to novelty stems not from the tradition alone, but also from the learning curve inherent in novelty. This is similar to the resistance of a modern-day user to shift onto a Dvorak keyboard from a QWERTY one. The layout of Dvorak might be more efficient, but the learning curve required to rewire the fingers from QWERTY on to Dvorak makes acceptance of novelty challenging.

As discussed previously in this chapter, certain pedagogical tools encourage dynamic interaction between tacit and explicit knowledge within the knowledge spiral proposed by Nonaka and Takeuchi (1995). These tools have been mapped on the knowledge spiral to explicate the relevance of each pedagogical tool to the knowledge transfer process (see Fig. 5). First, learning by observation is a socialisation process that enables learners to imbibe tacit knowledge, such as mental models and technical skills. Tacit knowledge can be shared in social interactions through "observation, imitation, practice, and participation in formal and informal communities" (Yeh et al. 2011, 147). Second, learning by emulating is the process of externalizing tacit knowledge into tangible artefacts through repetition. Third, learning by doing is the process of combining explicit experience of the master into the learner's explicit skills for making an artefact. Fourth, learning by reflecting is the internalization of the master's explicit knowledge into the learner's own heuristics of making. If learners are able to internalise their explicit knowledge, then they will be able to convert it into tacit knowledge (Hoe 2006). This leads to a cyclic learning model similar to Fig. 5 instead of a linear learning model.

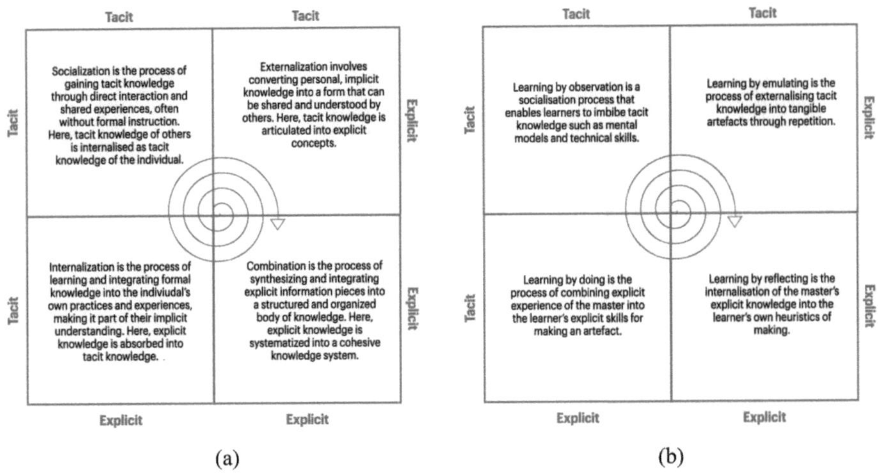

Fig. 5 **a** Knowledge spiral of Nonaka and Takeuchi (1995) **b** Pedagogical tools mapped on the knowledge spiral of Nonaka and Takeuchi. *Image source* **a** Recreated by the author from the knowledge spiral proposed by Nonaka and Takeuchi (1995). It was also published by the author in Patil (2025). **b** Author; it was also published by the author in Patil (2025)

Immersive and authentic work experience in making a variety of toys and collaborating with toy designers of varying skill sets pushes learners to acquire contextual problem-solving and decision-making skills that are necessary for responding to unpredictable production schedules and implementing design changes in response to market trends. Learners must discover by an intelligent effort how to integrate the parts of a whole that is greater than the sum of its parts—a skill that is not acquired by learning and performing in fragments but by practicing the knack of coordinating them effectively (Ainley 1993, 13).

6 Method of Evaluation Based on Risk

> Every tailor who described his apprenticeship to me included an account of the day his master left the shop in the middle of sewing a pair of trousers for a customer and told him to finish it. The test indicated the master's willingness to have the learner's work treated as though it was produced by the master, for sale by the master in the shop. (Lave 2011, 79)

Here, Lave describes a distinct point of departure from apprenticeship into the master's shoes in the community of Vai and Gola tailors. Failure in the informal, impromptu test not only demotes the learner but also results into a loss of revenue and possibly a loss of credibility for the master. Success in the test results in a new status for the learner in which he stitches a variety of garments in response to client demand, and rarely responds to his personal interests and playful impulses. In every new design, the newly minted master is taking a risk, but he is also learning and earning at the same time.

The risk taken by toy designers in executing various tasks to make an acceptable, and sometimes outstanding, prototype is embedded in the learning pedagogies, as discussed previously. The nature of the toy-making practice exposes learners to the principles of the artisan approach to risk from the very beginning of the learning cycle: observation. Learning to *see* is an essential skill set that learners in cultural design communities are encouraged to develop in childhood itself, and continues to be used by mature cultural designers as well. Emulating cultural objects and processes involves low levels of risk that can be mitigated by rigorous practice and astute observation. New learners in the *Bāṇārasī* community are often introduced to typical, familiar, simple, high-selling, and popular small items such as spinning toys or a *Hilāntā*. These typical toys have been mastered by mature toy designers, who are therefore adept at correcting learners' errors. Given the expertise of the masters over such typical products, the workmanship is based on certainty. Another set of toys made by the community involves making unfamiliar designs with well-practiced skills. The skill set is known, so the risk of failure is moderate. Yet observation and emulation alone are not enough. Intelligent iterations through constant prototyping and reflection upon failures allow toy designers to reduce the risk of making an unfamiliar and unpracticed design.

Learning by doing is often used in design education to represent, externalise, or replicate the thought process. This limits hand work to the workmanship of certainty, which is pre-determined and lacks decision making or risk in its operation (Pye 1968, 20). On the other hand, thinking-enabled production, as observed in *Bāṇārasī* toy design, explores problem solving on the factory floor itself. This practice of learning and upgrading, which is commonly found in cultural design pedagogy, has not yet been explored to its full advantage in design education.

The maximum risk is taken by cultural designers when they make an unfamiliar design in an unfamiliar process. The risk is even higher if the material is new. For instance, wood carvers who sculpt figurines were asked by a client to carve details of a figure on a lathe-turned surface. According to the client, the style of the carver was very different from what they were thinking, but after seeing the carvers' work, they decided to let them "go crazy creating one from scratch" (as quoted by Dame et al. 2011). No details of the carving were provided, and there was no similarity between the 2D design and anything the carver had made before. The carver not only had to imagine the character without any reference but was also unaware of what the client expected. The risk of failure was high (see Fig. 6).

Consider the example of the Tharu basket weavers who reside in the Valmiki Reserve Forest located in the Terai region of Uttar Pradesh, India. Women in the tribe are familiar with and skilled at making circular baskets with colourful motifs. They have learned the skill from their mothers and are passing it on to their daughters and daughters-in-law. Like the wooden toy designers of *Bāṇāras*, the daughters and daughters-in-law of the Tharu community also undergo apprenticeship under the senior women and learn by observation, emulation, doing, and reflecting. Learning, of course, has its own risks, as discussed previously. However, the master basket

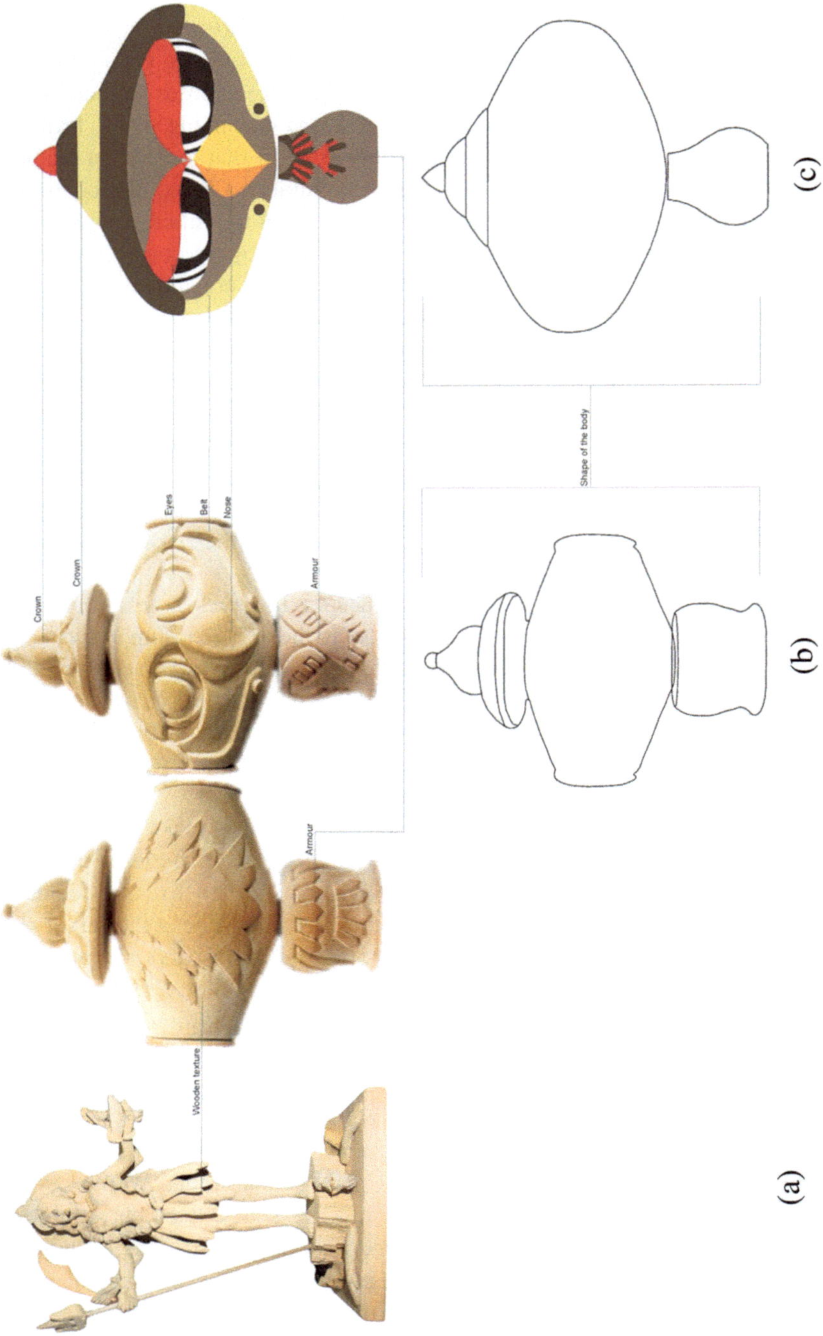

Fig. 6 **a** Existing Bāṇārasī design **b** new Bāṇārasī design based on the client-design **c** client-design. *Image source* Author

Fig. 7 Trajectory from typical to atypical. *Image source* Author

weavers themselves also take the risk of learning a new design. Despite being highly skilled in the cultural skill of basket weaving, the women took several attempts to shift their product category from baskets to furniture. They failed often in understanding the new scale (dimensions) of seating, as well as the load requirement of the new category. However, they did learn to reduce the risks enough to make a low-fidelity prototype at the end of a training workshop held in 2016. The risk of failure increases even more when we introduce an unknown object, such as a bean bag, which the makers are to fashion with the same grass and technique as used to create the baskets. The farther one moves away from the typical members of the category, the higher the risk of failure. Success in making an atypical product requires higher cognitive load and physical aptness than the practice of creating a typical product. Atypical products require a longer learning curve than typical products (see Fig. 7).

Each cultural design practice offers a range of physical, cognitive, organizational, and dispositional learning skill sets unique to its own worldview. This makes the learning pedagogy of every community of practice contextual and allows for specific "signature pedagogies" (Chick et al. 2009). Hence, it is important to assess each cultural design practice for propensity toward taking risk and the tools and methods available to evaluate risk. Risk evaluation is a challenging task because it requires the mapping of learning sequences as well as several different scopes within which learning processes are organized and accessed. Lave (2011) evaluated the transfer of knowledge and skills through such signature pedagogies (Shulman 2005)by studying the transfer of knowledge from one situation to another situation (for example, transferring mathematical skills that are required in tailoring to other situations beyond the realm of tailoring, such as pure or formal mathematics). However, Lave did not evaluate the risk involved in situational learning itself. Of course, in contextual pedagogies of cultural learning systems, sale of an artefact provides a general criterion for evaluation. But errors in artefact construction are more specific and translate into risks of monetary loss. The awareness of such potential losses should spur the design community and cultural designers to maintain high standards of good and acceptable work.

If cultural design must be understood as an impulse to do a job well for its own sake, then it is essential to evaluate the work of cultural designers based on the risk they take in making the object rather than the market risk of sale. That is, the cultural designer may have taken immense risk of workmanship but may still fail to please the client; in this case, the effort taken to do the job well should still be valued and appreciated. Moreover, from a pedagogical point of view, the risk-taking capability of a student is a measure of the mastery gained by him at improvising and innovating a tradition. The mastery over material and process that Sennett (2008) observed in the works of Stradivari and others is a result of the workmanship of risk. Hence, the job of a master in cultural practices is not only to teach learners to follow a tradition but also to instill in them a sense of capability to handle risk that is acceptable within the community practice and at the same time adds novelty to the tradition.

References

Ainley MD (1993) Styles of engagement with learning: multidimensional assessment of their relationship with strategy use and school achievement. J Educ Psychol 85(3):395

Alexander C (1964) Notes on the synthesis of form. Harvard University Press

Bandura A (1986) Social foundations of thought and action. Englewood Cliffs, NJ 1986(23–28):2

Chick NL, Haynie A, Gurung RAR (2009) From generic to signature pedagogies: teaching disciplinary understandings. In: Exploring signature pedagogies. Routledge, pp 1–16

Dame C, Sheel M, Bourah S. O holy artisan: where art Thou? Retrieved from Little Known Heroes: http://blog.littleknownheroes.com/2011/06/o-holy-artisan-where-art-thou/. Accessed 20 Jun 2011

Hallnäs L, Redström J (2001) Slow technology—designing for reflection. Pers Ubiquit Comput 5:201–212

Hoe SL (2006) Tacit knowledge, Nonaka and Takeuchi SECI model and informal knowledge processes. Int J Organ Theory Behav 9(4):490–502

Jones JC (1970) Design methods: seeds of human futures. Wiley, New York

Lave J (2011) Apprenticeship in critical ethnographic practice. University of Chicago Press

Lawson B (2006) How designers think—the design process de-mystified. Architectural Press Publications, Oxford

Leadbeater C (2009) We-think: mass innovation, not mass production. Profile Books, London

Nonaka I, Takeuchi H (1995) The knowledge creating company. Oxford University Press, New York

Patil K, Athavankar U (2022) Design precedents to design innovation: category-based reasoning in problem-solving. Des J 25(1):62–85

Patil K (2025) Artisanship of risk: the uncertainty of traditional design. J Vocat Educ Train 1–22

Pye D (1968) The nature and art of workmanship. Cambridge University Press, Cambridge

Reitan JB (2006) Inuit vernacular design as a community of practice for learning. CoDesign 2(2):71–80

Schön D (1983) The reflective practitioner, how professionals think in action. Basic Books

Schön DA (1992) Designing as reflective conversation with the materials of a design situation. Knowl-Based Syst 5(1):3–14

Sennett R (2008) The craftsman. Yale University Press, New Haven

Shulman LS (2005) Signature pedagogies in the professions. Daedalus 134(3):52–59

Simon HA (1969) The sciences of the artificial. MIT Press, Cambridge, MA

Sullivan LH (1896) The tall office building artistically considered. Lippincott's Monthly Mag 57:403–409

Yeh Y-c, Huang L-y, Yeh Y-l (2011) Knowledge management in blended learning: Effects on professional development in creativity instruction. Comput Educ 56(1):146–156

Chapter 7
Thinking Like a Cultural Designer: Juggling Worldview over Livelihood

1 To Be (a Handmade) or Not to Be

At first glance, the worlds of Moroccan carpet weavers and French organic farmers seem disparate, even incongruous. The latter, some might argue, scarcely fit within the bounds of cultural design. However, through the discerning lens of historical contexts, embodied knowledge, and the transformative process of repetitive engagement with materials and tools, Myriem Naji (2020) brings these seemingly disparate practices into a shared frame. Naji's longitudinal ethnographic study reveals the resilience strategies both communities employ to withstand the relentless forces of global trade and production.

In the Sirwa region of Southern Morocco, female carpet weavers find themselves at the mercy of gender and geographical disadvantages within a male-dominated global carpet trade. Their marginalization is compounded by limited access to capital and markets, entrenched patriarchal power dynamics, and the invisible nature of their artisanship. This marginality is a position they yearn to escape. For many young women, securing a wealthy husband offers a way out—a deliberate submission to patriarchal norms, yet also a strategic escape. Failing that, they continue weaving within the confines of their homes, honing their skills through self-learning and continuous innovation to meet market demands, particularly in Marrakech. Their strategy hinges on supporting the men in their families, who broker their designs in urban markets, and they firmly believe that economic salvation lies solely in the hands of men.

In contrast, the organic farmers of Tarn in France embrace the laborious and financially modest practice of small-scale organic farming with pride and satisfaction. Many have forsaken comfortable urban jobs for a life harmoniously intertwined with nature. Eschewing mechanization, hybrid seeds, and chemical fertilizers, they meticulously cultivate their produce and advocate for its taste, nutrition, and freshness in local markets—an ideal of sustainable living. Rather than seeking to escape

K. Patil, *Toying with the Mind*, Design Science and Innovation,
https://doi.org/10.1007/978-981-96-9227-9_7

marginality, they find ways to celebrate it, and they view it as integral to their well-being.

Marginality can also be navigated by candidly embracing mainstream opportunities. The 2008 case of Sainik Farms in Morādābād, India—a cultural design community of brassware—illustrates this gentrification. The proprietor, upon his recent return from Europe, observed a market for bison skulls among European youth. Consequently, hundreds of bison skulls were cast in aluminium by Morādābād's brassware designers—an endeavour that diverged sharply from their generational Nakashi designs. These skulls, unsellable locally due to Islamic prohibitions against graven images, were produced solely for foreign markets to generate profit. Here, cultural designers prioritized livelihood over worldview and accepted commissions to survive rather than innovate like the Sirwa women. They neither took pride in their work nor maintained a staunch allegiance to tradition, instead succumbing to modernization's demands to secure a decent standard of living.

The handmade world oscillates between exalting the hand's identity and concealing it to mimic modern design. The former embodies the cultural design's soul, while the latter transforms it into a commercial product. This dichotomy underscores the struggle between maintaining tradition and adapting to modernity for survival. Adhering to tradition fosters a distinct identity but renders products less viable in contemporary markets. Conversely, abandoning tradition opens new livelihoods at the cost of community identity. This precarious balance prompts the question: Why do cultural designers grapple with this dilemma while modern designers do not?

2 The World and Its View

A worldview, simply put, is a cultural landscape—a lens through which people think, act, and interact with peers, materials, technologies, and their environment, and make, use, and reject objects. Ingold (1993, 153) refers to it as the "dwelling perspective"—a view steeped in the past. A worldview is an enduring record of previous generations who have inhabited it. These "lived-in worlds" form a communal knowledge pool shared by all members of a community of practice (Wenger 1999). These communities, self-organizing and defined not by tasks but by participation, engage in mutual problem solving through shared knowledge and practices, which guide members in designing better solutions.

The extent and nature of these lived-in worldviews vary between communities. Modern designers, for example, work in isolated studios, observing the "real world" through the artificial construct of design labs. Ethnographic methods attempt to bridge the gap between user needs and designer aspirations, yet users "remain others"—distant representations of reality. Cultural designers, however, often design for contexts they intimately understand; sometimes they are users of their products.

Bāṇārasi toy designers, for instance, create toys for religious festivals they themselves celebrate and also for their children, which makes them acutely aware of design flaws from a personal experience without needing external feedback.

Yet this is not universally true. A *Ṭāṇṭī*, or weaver, cannot afford the intricate *Jāmdānī* sarees he weaves. This process, which can take years, requires immense skill. The *Ṭāṇṭī*, sells these sarees to patrons while his wife wears simpler garments. Nonetheless, he understands the saree's function, social value, and the wearer's desires, which are informed by the daily life of the *Ṭāṇṭī* with the women in his household. Similar dynamics apply to exquisite cultural designs worldwide, which are designed for specific patrons but rooted in the designer's lived experience. Here, living and user spaces merge, which allows for immediate feedback on experimentation and risk.

The cultural designer's worldview shapes design decisions that range from not changing a design (unchanging) to accomodating change that is assimilated over time (Hubka 1979). This adherence, which is often seen as a limitation, is a boundary within which responsible innovation occurs—innovation that adds value and meaning. Modern design, in contrast, often disrupts worldviews in the quest for innovation, thereby embodying Ezra Pound's (1934) "make it new" mantra. In cultural design, the tacit knowledge embedded in practice resists codification, unlike modern design, which seeks to translate the tacit into explicit knowledge.

Cultural designers often avoid formalizing their practice. Modern design emphasizes explicit knowledge, while cultural design retains tacit knowledge. These tacit codes, though invisible in theory, are evident in how cultural designers organize tasks, display objects, and engage with their community. Vernacular methods of categorizing objects offer insights into the everyday folk theory of categories (Lakoff 2008, 5), shaping how cultural designers retain, modify, or introduce characteristics in their work. Unlike modern design, which operates on the periphery, cultural designers often aim to create prototypical forms that celebrate the persistence of their identity. Cultural designers may be criticized for a lack of innovation, yet even their most ardent detractors cannot ignore their profound sense of identity. To create a product that embodies the essence of a community's identity, one must be thoroughly acquainted with the community's conventions, canons, and customs of making, consuming, and innovating designs. For instance, the *Bāṇārasi* generational methods of designing a *Khilōnā* at the intersection of the conventions, canons, and customs are a testament to the physical manifestation of *Bāṇārasipan*, or *Bāṇārasi*-ness. *Bāṇārasipan includes the* inherent characteristics of the toys that make them distinctly identifiable as a *Bāṇārasi Khilōnā*; it captures the essence of what makes the toy "*Bāṇārasi*-like" rather than just a concept or idea. However, *Bāṇārasipan* is not only perpetuated through the generational persistence of the same motifs, but interestingly is also fueled by features and sometimes entire artefacts adopted from external cultures. These unfamiliar features and artifacts extend beyond the established *Bāṇārasi* categories, but the toy designers not only emulate but also modify borrowed features to resonate with their traditions. The unfamiliarity of these elements is often balanced by their strategic coupling with familiar design precedents. In situations involving foreign artefacts or ambiguous category boundaries, some features are appropriated or reverse engineered with minimal alteration, while the predominance of familiar characteristics ensures coherence within the designers' worldview. Consequently, these assimilated features, aligned with established conventions and customs, ultimately fit within the accepted continuities of toy design.

Such intercultural exchanges have consistently driven change within insular communities, even during periods of limited interaction. Despite a rigid worldview that often discourages deviations from established norms, cultural design communities frequently incorporate elements from other cultures. Lynn Stephen's (1991) analysis of four indigenous groups—the *Zapotec* weavers of *Teotitlán de Valle*, Oaxaca; the weavers of Otavalo, Ecuador; the *Nahuatl Amate* bark painters of Guerrero, Mexico; and the *Kuna mola* makers of San Blas, Panama—demonstrates how these cultural design communities maintained and even strengthened their identities amid pressures to assimilate. Similar to the *Bāṇārasi* toy designers, these communities differentiate between products intended for various audiences, including tourists, importers, and their own community members. Indigenous identity is often commodified for cultural consumption and involves reclassifying objects. For example, *Zapotec* weavers integrate cultural motifs and Navajo designs into their textiles and create pieces inspired by modern artists like Miro, Picasso, and Escher for the American market. Similarly, Oaxacan artists depict ancestral figures and deities in their *amate* bark paintings alongside imagery related to the U.S. border and the term *migra*, which refers to border patrol agents. The *Kuna* community exemplifies a strategy of recategorization rather than assimilation. Cultural *molas* worn by *Kuna* women have been transformed into fashion items marketed directly to tourists or through cooperative networks in urban markets across America and Europe.

However, willful borrowing of features and their adaptation is not the market choice often given to the *Bāṇārasi* toy designers, especially in client orders, which are often disruptive or disrespectful of their worldview. Knowingly or unknowingly, the *Bāṇārasi* toy designers have formed a strategy to preserve their identity while still surviving the modern industrial onslaught. To earn their livelihood, the toy designers have proficiently learned the skill of making modern artefacts or packaging the *Bāṇārasi* identity for consumption outside *Bāṇāras*. *Decoration ka Khilōnās* are not designed for *Bāṇārasi* patrons. They are painted "*fancy*" to attract modern customers and often flout the *Bāṇārasi* canons, conventions, and customs, thereby exploiting the ignorance of the consumers of the community worldview and hence that of an authentic *Bāṇārasi Khilōnā*. But despite the requirements of customer orders, they steadfastly refuse to incorporate features unacceptable to their worldview into the *Bāṇārasi Khilōnās*; in this way, they use themselves to protect, preserve, and pursue their *Bāṇārasipan* or *Bāṇārasi*-ness. Preservation of the cultural cues of *Bāṇārasipan* in the *Bāṇārasi Khilōnā* designed for local patrons is a strategic, though implicitly agreed upon, decision of the community to position themselves as handcrafted, traditional, authentic, or exotic against the machine-made products that modern customers seek. Notably, the toy designers have reached a consensus on this strategy despite the absence of centralized leadership and lack of explicit methods of making collective strategic decisions.

In this sense, cultural design is a source of livelihood and a worldview that scaffolds meaning making, understanding, and interacting with the world. It is a creative urge manifested in the living tradition of making, representing, and forming propinquities via the products. Consequently, the products, the process, and their proximity zone with users represent a microcosmic view of the larger worldview. Such worldview-driven practices are not easily encapsulated in the English words *craft*, *art*, or *design*. Stella Kramrisch (1958) contends that the Indian word *Śilpa* cannot be truly translated into English because it indexes several signified: art, skill, craft, labor, ingenuity, rite and ritual, form, and creation. It is not an ordinary parity product made by designers detached from the local circumstances in which it is used, made meaningful, and valued. For instance, the art and design of making an *Aṣṭadhātu Lallā* and *Lallī* (baby *Ram* and *Sita*) in Ayodhya, India, is not merely a skill of practicing the lost wax process but also a ritualistic performance of manifesting the divine in a humble material molded by human hands. This ritualistic aspect can also be observed when the carpenter prays to the tree for forgiveness before cutting it for wood. Verses 57.10-11 of the *Bṛhat Saṁhitā* describe such a prayer seeking forgiveness from a tree:

अर्चार्थममुकस्यत्वंदेवस्यपरिकल्पितः। नमस्तेवृक्षपूजेयंविधिवत्सम्प्रगृह्यताम्॥

यानीहभूतानिवसन्तितानिबलिंगृहीत्वाविधिवत्प्रयुक्तम्।

अन्यत्रवासंपरिकल्प्यन्तुक्षमन्तुतान्यद्यनमोअस्तुतेभ्यः॥

Oh, Tree, you have been selected for the worship of the particular deity. Salutation to you! I worship you in accordance with the rules; kindly accept it. May all those beings that dwell in this tree fix their residence elsewhere after receiving the offerings made according to the rules. May they forgive us now; We bow to them.

The ritual is not only made explicit in the prayer but also in the practice of conservation. The marginalized households in the Indian state of Uttarakhand have traditionally saved three trees for the lifetime of each family member. They cut one tree when a child is born to celebrate and save money for caregiving; another when the child grows up, gets married, and starts a family; and a third to cremate the old body of the dead. The community does not feel the urge to cut more trees than required. Therefore, the indiscriminate felling of trees to the Amazon rainforest is incomprehensible in the worldview of a common mountain dweller in Uttarakhand.

Such ways of life and production were generationally consumed by patrons who shared the context of the producers in the same time and space. The patron was not only a part of the financial transaction of the produce but also a stakeholder in the production of worldview-appropriate and acceptable cultural designs. For instance, the unwritten but implicitly followed *Bāṇārasi* conventions, canons, and customs are known to the producers and, to a large extent, the patrons, who can reject inappropriate designs. Design decisions regarding the semantics of the *Bāṇārasi* toys are strongly governed by worldview-approved customs, which may be religious or social. The influence of the customs is so strong that they are far more numerous than the conventions and canons in the cultural pool of design rules and are rigidly respected by designers and patrons. Sometimes, even a foreign or urban customer may reject a *Khilōnā* because it violates a custom. Ajit Kumar Viśvakarma placed Lord *Gaṇeśa* on a peacock instead of on the worldview-approved mouse. Even foreign customers refused to purchase the irreverent design, which did not reflect the local religious beliefs. This implies that the cultural design sector is value driven. Customs allude to the ritualistic, religious, and cultural norms of the community by which the people live and measure the goodness of a person or object. Therefore, a *Bāṇārasi Khilōnā* should not only look pretty and provide ludic satisfaction but also adhere to the value system of the community (Patil 2017). Consequently, Bāṇārasi toddlers are not given Barbie dolls to play with, not because of the exhorbitant price or the industrial process and unmsustainable materials, but because the dolls are not *Sabhya*. *Sabhya* indicates the cultural notion of appropriate dressing among women and girls, who are expected to be fully clothed, have coy eyes, have a polite demeanor, and exhibit cultural signifiers such as *bindi*, *pallu*, and bangles. It is the worldview-approved notion of goodness in the role model given to a toddler. The culturally puritanical prototype of *Sabhya* would be *Radha*, the consort of Lord *Kṛṣṇa*, who is represented wearing a saree, with a *pallu* drawn on her head, a *bindi*, earrings, bangles, and *kamalnayani* eyes. Her body is not slender and curvy like the Barbie doll. The shape of the doll is indicative of human body parts and not sensuous in the least. *Sabhya* cannot exemplify a body shape that promotes the sexual objectification of a woman's body, as does the Barbie doll (Ward et al. 2023). Surprisingly, despite knowing the effect Barbie has on the body image of young girls, Mattel continues to manufacture Barbies, and modern parents continue to consume them for their children. The design worldview that promotes the consumption of Barbie and the commodification of a female body is a good example of the modern designer, whom Papanek and Fuller (1972, ix) have accused of putting "murder on a mass-production basis. By designing criminally unsafe automobiles that kill or maim, by creating a whole new species of permanent garbage to clutter up the landscape, and by choosing materials and processes that pollute the air we breathe, designers have become a dangerous breed."

Unlike modern design practice, the acceptance of an artefact in the *Bāṇārasi* community is independent of novelty and rather dependent on its appropriateness within the community-held notions of propriety; adherence to the worldview-sanctioned authority; and adherence to the convention, canons, and customs of the practice and affinity with the generational designs (Patil and Athavankar 2012). Deviation or disruption of the worldview is not a norm in cultural design practices. In the *Bāṇārasi* community, toy designers, shopkeepers, and patrons do not evaluate toys on novelty but on their acceptability in the kinship of community-owned objects. The worldview acts like a self-regulatory mechanism that quickly discourages inappropriate changes in the cultural designs and moderates the acceptability of artefacts and elements from outside the community.

However, these self-regulated and self-organized systems are constantly fighting a battle in which sometimes they lose their livelihood for survival. The historical narrative of loss built around these communities against colonial, capitalist, and other exploitative or opportunistic regimes has categorized them as vulnerable, perishable, marginalized, and victimized. However, there is also a narrative of survival, resilience, and strategy that goes unnoticed if we ignore the intentional persistence of their worldview. For instance, it is well known that heavy extraction of raw material from the Indian hinterland to the British textile industry in Manchester and Leeds broke the back of the Indian handloom sector. The story of this unfair trade continues, as weavers in India even today commit suicide due to a loss of market to the machine-made cloth. However, Tirthankar Roy considers that "by the end of the nineteenth century, not only had a large number of weavers survived competitions from the British cloth, they were also investing in new tools and processes" (2002, 507–508). How did several hundred thousand illiterate, unorganized handloom weavers escape the well-planned, rigorously executed strategy of the highly educated English *Saheb*? There must be something that we are missing about the survival instincts and strategies inherited by cultural communities.

3 Necessity of Interested Interests in Cultural Design

In examining the interplay between cultural justification and economic motivation, we uncover a landscape where the very fabric of cultural design communities is often frayed by market forces. This dynamic interplay, frequently orchestrated by well-meaning designers, NGOs, and policymakers frames cultural designers as passive subjects in need of innovation. In the process of legitimizing the participation or intervention of modern designers in cultural design practices, the identity of the cultural designers as problem solvers, managers, and strategists has often been erased and redrawn as a primitive practice—an art form deemed out of step with the present and in need of scaffolding for revival and preservation. At the heart of this narrative lies a troubling assumption: that cultural designers are not resilient enough to negotiate unfamiliar modern markets, reconstruct their identity outside the local, assert their voices, or adapt to changing circumstances, technology, and user patterns to steer the course of their survival. On the contrary, DeNicola (2005, 251–18) points out that in some instances, well-meaning interventions of designers and other experts limit the opportunities for the cultural community to become modern and innovative on their own terms.

The gentrification of cultural design by designers, policymakers, and NGOs in a quest to uplift, protect, and modernize a dying industry has often forced cultural communities to innovate or invent outside the permissible boundaries of their worldviews. The often well-meaning interventions proposed and implemented by different agencies have not been accepted in the community worldview, and the results are disheartening. Some of these interventions in India are briefly discussed below to illustrate the different levels of resistance offered by the communities.

(a) The *Bāṇārasi* toy designers, skilled designers of nested dolls, turn a plethora of shapes and motifs to fulfill client orders. Among these motifs, one finds representations of girls with short hair, makeup, curvaceous forms, and short dresses. Such depictions diverge significantly from the earlier discussed ideal of the *Sabhya gudiya* within the *Bāṇārasi* cultural context. Yet, despite this discordance, the toy designers persist in creating these figures, driven by the exigencies of their livelihood. However, this persistence is not born of appreciation. The toy designers themselves disdain these sensuous proportions, and deem them morally unsuitable. They take care to shield their own daughters from these dolls because they are wary of potential moral degradation. But this protective stance extends to their children alone and demonstrates a clear boundary: while they do not subject their offspring to these figures, they remain indifferent to the moral implications for children of other cultures who encounter these creations. Thus, the *Bāṇārasi* toy designers do not intervene in the cultural paradigms of others, irrespective of perceived inappropriateness. They maintain a firm resistance to external influences, safeguarding their cultural integrity by keeping the practice of their cultural design within strictly defined boundaries. This boundary setting underscores a tacit acceptance of economic necessity while simultaneously upholding a moral order within their own community.

(b) In the bustling lanes of *Vishwanath Gali*, shopkeepers act as intermediaries between the English-speaking urban clientele and the *Bāṇārasi* toy designers. Yet this mediation is not without its frictions. The shopkeepers often voice their frustrations, lamenting the toy the apostrophe on designer‘ seeming disregard for their instructions. They note instances where toy designers alter specifications for convenience, miss deadlines despite being aware of them, take shortcuts, and show a marked lack of urgency when market demand is low, all of which further complicate transactions. This behavior places the shopkeepers in a precarious position and exposes them to the ire and occasional humiliation of their clients.

A particular point of contention is the refusal of the *Bāṇārasi* toy designers to eliminate the red blotch on the wings of the wooden birds they produce. These red blotches clash with the modernist designs favored by many clients (refer to Fig. 1). Despite clear instructions to forgo this detail, numerous toy designers persist in its use, much to the frustration of the shopkeepers. This resistance serves as a subtle assertion of the toy designers' cultural worldview and defiance against the market's relentless push to alter their aesthetic sensibilities.

The relationship between shopkeepers, clients, and toy designers is marked by a continual testing of boundaries, yet these boundaries are rarely broken. Toy designers strive to minimize material usage, reduce effort, and expedite turnaround times, all while seeking the highest possible profit. Conversely, shopkeepers and clients are in pursuit of toy designers who can provide high-quality materials, exhibit exceptional artisanship, invest significant effort and time, and produce standardized and precise toys at minimal manufacturing costs. This dynamic creates a delicate balance, a tightrope walk where each party seeks to maximize their gains without severing the economic ties that sustain their livelihoods.

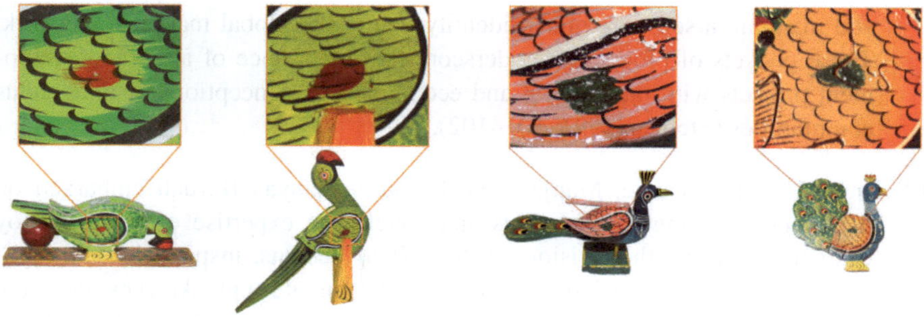

Fig. 1 Red blotch on the wings of birds, often perceived as an eyesore by modern customers. *Image source* Author; also, published by the author in Patil and Athavankar (2025)

(c) The *Bāṇārasi* community of toy design, which is deeply embedded in a tradition of independent production, has long nourished both the cultural and economic fabric of the society. This cultural design is not merely an economic activity but is interwoven with a rich cultural framework that encompasses institutions and practices such as reciprocal labor exchange, skill learning, and a network of rituals integral to community life. Within this insular world, these practices sustain a collective identity and a sense of continuity.

However, when these cultural designs are projected onto the external world, their essence is often stripped away and reduced to mere commodities for economic transactions. The ethnic identity of the *Bāṇārasi* toy designers as perceived by outsiders is shaped largely by market forces and devoid of its cultural roots. These identities are curated by market agencies, whose experience in promoting sales eclipses the voices of the toy designers themselves. This commercial appropriation frequently forces communities to alter their cultural designs and creates dissonance with their worldview.

In response to these pressures, the *Bāṇārasi* toy designers have developed nuanced strategies to safeguard their cultural identity while engaging with the capitalist market. They have created distinct categories to differentiate their authentic cultural products from those designed for external consumption. For instance, *Decoration ka Khilōnās* are designed with greater precision and detail and feature an *English* pastel color palette as opposed to the vibrant, contrasting *Chatak* colors typical of *Bāṇārasi* traditions. This distinction sets them apart from religious *moortis* and ludic *khelne ka Khilōnās*, or toys for play, thus maintaining a clear boundary between cultural authenticity and market-driven adaptations.

This strategic categorization allows the *Bāṇārasi* toy designers to navigate the demands of tourism and export markets without compromising their cultural integrity. By creating a separate category for non-*Bāṇārasi* styles, cultural designers enhance their livelihood opportunities while preserving their cultural values. Such adaptive strategies offer valuable insights for development planners and demonstrate how communities can achieve economic self-management and maintain a self-determined identity within the global market framework. These pockets of production underscore the importance of aligning development projects with the cultural and economic self-conceptions of indigenous communities (Stephen 1991, 101–102).

(d) In 2011, Chris Dame, Manjari Sheel, and Soujanyaa Boruah embarked on a project involving wooden toys and sought the expertise of *Bāṇārasi* toy designers to bring their visions to life. Their designs, inspired by anthropomorphic warriors from Indian mythology, began as rough sketches and were refined using software to finalize the forms, colors, and patterns. These finalized designs were handed to the *Bāṇārasi* toy designers with the expectation of precise replication in wood. The results were mixed. They noted the following:

The one on the left is the closest copy of our original Jatayu we have seen, which is very promising. The claws on the front are horrible, though. The middle one had a gorgeous texture that is very typical of the *Bāṇāras* style all over the top. It's definitely a keeper. The Vasuki on the right, though, is the worst thing we've seen by far. It looks like a little kid did it. Apparently, the craftsman spent all of his time on the first two and rushed through the last one because he ran out of time. In addition, the high gloss they added makes the toys look really cheap. Not exactly the best impression, and we need to talk about it if we want to give this guy a second chance or go with the exporters we spent all day with.

The modern designers expected the cultural toy designers to execute their designs and not cogitate over the final design because they were unaware of the customers and lacked the expertise to design a modern toy. This attitude reflects a broader pattern, in which designers from urban, English-speaking backgrounds often engage with *Bāṇārasi* toy designers in an effort to bridge cultural artisanship and modern design. These engagements often take the form of company projects, institutional development initiatives, Corporate Social Responsibility (CSR) activities, or academic assignments.

These interventions or assistance often reinforce the preconceived division between manual versus mental labour, wherein modern designers are the thinkers and the cultural designers merely implement the plan of the former with little or no creative input (Balaram 1998, 2005). This is highly ironic considering that the erstwhile designers—that is, the cultural designers—have created and innovated for centuries without any formal education or technologically advanced tools like Computer Numerical Controls (CNCs), Rapid Prototyping Technologies (RPTs), or Artificial Intelligence (AI) to aid their creativity and their hand skills. Today, however, innovation, creativity, planning, strategizing, and user empathy largely form the repertoire of English-speaking, formally educated, often urban-based graduates of design schools, who heavily rely on external aids like mood boards or AI for drawing a solution.

Almost all cultural designers accept such orders even though it downgrades their status to a technical labour class. Livelihood is crucial for their everyday survival, and they understand the cash flow enabled by such mindless labor. They do not resist such work, and they complete it to the extent possible. However, their acceptance to complete a task does not imply their consent for adopting such work into their worldview. They are not enamoured with the varied styles they are exposed to by urban clients. If a feature is adopted, it is assimilated and not merely mimicked to fit in the worldview. In all likelihood, it's a tit-for-tat approach. All they offer the client is a skill while their worldview remains untouched by foreign influence.

(e) The Rampur *chakus* or knives were renowned all over India for their durability, ergonomics, sharpness, and elegance. They were produced locally by royal bladesmiths for the Nawabs of Rampur starting in the eighteenth century, after firearms were introduced. Their popularity even made them an enviable accessory of Bollywood attire. However, in the mid-1990s, the State Government of Uttar Pradesh banned knives longer than 4.5 inches (11 cm) of blade length. This reduced the market reach of locally produced knives, which were quickly replaced by cheap Chinese alternatives. Today, the local industry in Rampur has almost closed down. During my visit in 2010 to Rampur, there were only a handful of families selling small knives in the local market. A policy decision probably introduced to reduce the availability of unlicensed arms in the hands of the public led to the loss of the livelihood of an entire community and the death of a generational practice and know-how. Ironically, long Chinese blades replaced local production in the Indian market, making the policy ineffective for all practical purposes. The policy killed an entire practice of designing tools for offensive and defensive combat and for operations outside of combat, such as agriculture. The invisible death blow to the community practice was not noticed by anyone. The hands of the state were so strong that, interestingly, the very community that facilitated combat could not defend itself against the policy that caused its slow death.

4 Negotiating Interests of the Interested in the Interesting

Since the Industrial Revolution, cultural design communities worldwide have resisted the onslaught of machine-made goods, decontextualized design processes, profit-focused designers, and consumers driven by immediate gratification. The resistance continues against all odds, and several champions of cultural design have emerged worldwide. In India, support for the handmade emerged in the *Swadeshi* movement in the pre-independence era itself, not in terms of design development or market analysis and branding but in nudging consumers to boycott foreign goods. Increasing consumption fueled the already well-oiled crafting machinery of the cultural communities. However, post-independence, the nature of these interventions transformed into scaffolding, hand holding, directing, envisioning, planning, and analysing on behalf of the cultural communities, which were presumably in dire need of external assistance. These negotiations have increased the contribution of the cultural design sector to the Indian GDP (gross domestic product). However, they have not necessarily created an empowered and independent population capable of making design decisions for themselves. The parameters of success defined for cultural design communities by designers are inadequately described in terms of revenue alone; they should also encompass means of production, learning, social mobility, and design thinking—in short, they should enrich the designers' worldview. Anthropologist Lynn Stephen, in a 1991 study, highlighted that the success of a cultural design community is characterized by the degree of control producers maintain over the marketing and distribution of their goods. Furthermore, their authority to decide how much of this income is reinvested into community institutions predicts the sustainability and viability of their small-scale ventures. It is crucial not only to utilize labor in production but also to have the freedom to reinvest it into community rituals and political institutions. This perpetuates and replicates institutions and events that scaffold the formation of local ethnic identity. For resilience against market vagaries and climate adversities, Stephen also advocated integrating cultural design production with agricultural and allied activities.

Additionally, my engagement with the *Bāṇārasi* community of toy designers, in particular, and other cultural designers of Uttar Pradesh, India, in general, gave me more insights into the success and failure of participative efforts of modern interventions in the cultural design communities. The condition of equitable participation in any participative effort is non-negotiable, often emphasized, but rarely implemented. Equitable participation is impervious to the inclusion of cultural designers as skilled technical partners for externalizing the design and planning processes of modern designers. Instead, they, too, are active cognitive collaborators in the problem-solving process. It is only when the condition of equitable participation is met that several culturally relevant parameters kick in. For instance, for a sustainable economic model, it is important that products are not only sold outside the community, which offers a larger profit margin, but also in local markets.

Local markets are an essential part of the ecosystem of a cultural design community; local markets enhance the community's resilience in dealing with the vagaries of the market, establish a local requirement for nurturing the cultural design units, and cultivate the design worldview of the community. Only through a firmly established local ecosystem do the novel products and processes introduced through the interventions of the modern market become a part of the cycles of emulation between adult design peers and between the design master and apprentice. If a product or process is not emulated beyond the requirements of the client order, and if recording devices are lacking, the community is unable to memorise the template and, hence, is unable to produce the design again of their own volition. Consequently, the innovation introduced by the intervention dies when the client order is completed. Emulation, specifically the emulation of apprentices of their masters, is the most reliable measure of the continuance of an innovation. An emulated product is often christened in the local dialect, which is the highest level of assimilation. Products and processes borrowed from other cultures continue to be "othered" within the receiving culture until the receiving culture accepts it enough to include it in its language or local parlance. *Chugtī Chidīyā, Hilāntā,* and *SukhīParivār* are instances of designs borrowed from client orders by the *Bāṇārasi* toy designers. These designs have been heavily emulated by the community members for purposes of learning and earning and have been linguistically christened in the local dialect, which confirms their complete acceptance in the community worldview. Such local parameters of acceptance of an innovation or design in the cultural community enhance the chances of the survival of newness in a cultural design practice, which is often viewed as rigid and unchanging.

Protecting the worldview comes at the cost of livelihood and is often considered a limitation to the creativity of cultural designers. However, this study discusses this limitation as a context like any other, which has shaped some understated but evolutionarily fit methods of learning and the production and evaluation of design decision in the absence of a leader. Walking the tightrope between the worldview and livelihood is a major dilemma facing the handmade sector. Therefore, in conclusion, this chapter will offer a few heuristics for equitable participation in the worldview of a cultural design community.

Heuristic 1: Assimilative processes are self-paced, more inclusive, and acceptable.: The thrust for newness in modern design forces all problem solving to generate disruptive solutions and denies visibility to any antecedent or preceding exemplars. The very premise of the modernist movement was to sever its ties with the traditional. Anything that is not tied to the past is modern in its approach and representation.

However, the worldview-driven practices of cultural design celebrate and respect their precedents as evolutionarily fit exemplars of problem solving and humbly emulating these models. Innovation, in this context, is not a disruptive force but an evolution of the past and is accelerated by the interaction of cultural practices with other cultural practices, both modern and otherwise. These other cultural practices encompass:

1. Other categories of products within the cultural practice itself
2. Products from other practices within the larger community in which the cultural practice is situated
3. Products from practices beyond the known range of the cultural practice.

The cultural design process acknowledges its roots while remaining open to influences contributing to its growth and adaptation, thereby ensuring that each innovation is a continuation rather than a departure from its storied past. For instance, the *Bāṇārasi* community of toy designers can borrow (a) products from the *Decoration ka Khilōnā* category to reconceptualise a toy from the *Khelne ka Khilōnā* category (that is, between-category exchange of elements), (b) elements of the *Bāṇārasi* silk sarees to create new patterns in the *Bāṇārasi* toys (that is, exchange outside the tradition of toy making but within the larger *Bāṇārasi* community, or (c) borrow an entire design or its elements from the Russian practice of toy making into its own (that is, exchange with an unknown cultural practice outside the geographical and social boundaries of *Bāṇāras*).

In several instances where new designs have been accepted and emulated by the *Bāṇārasi* community of toy designers, it was observed that the designers actively participated in the decision-making process, blending elements of their cultural practice with the designs from client orders or borrowed artefacts. Such assimilative practices, though not disruptive, facilitated a greater degree of change than the natural evolutionary process. This active engagement ensured that innovation was harmoniously integrated with the generational habits and allowed for significant adaptation while preserving the essence of the cultural design. Without being radical, the assimilative process was more inclusive, was fueled a moderate extent of innovation, was acceptable to the community worldview, and was willfully retained and emulated by the toy designers in their toy-making practice. Assimilative practices had a higher chance of retention in the community than disruptive interventions, which lasted only until the completion of the client order. Moreover, assimilative practices gave a voice to the community by democratizing the design process, engaging the community in a meaningful dialogue, and reinventing cultural forms of knowledge with the consent of its owners. Assimilative practices did not reduce the cultural designers into a skilled labour force executing the ideas of the modern designers. Therefore, collaborations with cultural practices should aim at assimilating the cultural practice with modern elements instead of subverting the cultural worldview with a disruptive solution regardless of how close the solution may be to the market demand, customer taste, or the designers' worldview.

Heuristic 2: Language is a tool for understanding a worldview. Lack of understanding of the cultural worldview has severely limited collaboration with and participation of well-meaning modern agencies and actors with the cultural design communities. Cultural knowledge systems are often different from and sometimes even antithetical to the modern view of things. This does not necessarily mean that they are outdated and irrelevant, but simply that there is a different cognition at the root. For instance, it would be incorrect to say that the Kuuk Thaayorre tribe from Pormpuraaw cannot understand the concept of left and right. Rather, their language exclusively uses absolute cardinal directions—north, south, east, and west—even for the smallest of scales. For instance, instead of saying, "The fork is on the left side of the plate," they would say, "The fork is in the southwest of the plate." According to Stephen C. Levinson of the Max Planck Institute for Psycholinguistics in Nijmegen, the Netherlands, and John B. Haviland of the University of California, San Diego (1994), people who emphasize absolute directions in their language possess exceptionally strong spatial cognition. Therefore, if members of the Kuuk Thaayorre tribe are placed in an unfamiliar, complex maze, they are far more likely to find their way out successfully compared to individuals who rely on relative directions. Their impressive spatial abilities challenge the limitations proposed by cognitive science regarding human navigation and orientation.

It is possible to understand such a different and uniquely embodied sense of being through the speaker's language. Of course, language is an entry-level barrier for most modern designers who want to gain the confidence of a cultural community. However, one of the primary requirements for "going native" is to speak the language of the community. Obviously, it may not always be possible to break this barrier, especially for a short duration of time. However, it is always possible to translate conversations or meanings of linguistic terms to some extent if we recognize that the meaning lies in the context and not in its translation, which may be an inadequate interpretation.

Unlike customs and conventions that require considerable time to be observed and interpreted in the context of their use, language is a direct and easily accessible tool for understanding the worldview of a community for appropriate intervention in the worldview. Language gives access to proper nouns, adjectives, and verbs related to artefacts and processes that are a reliable entry point into a community. Moreover, christening a borrowed feature, object, or process in the local dialect is the indicator of complete acceptance in a cultural community and hence the completion of the assimilation process. *Bāṇārasi* equivalents of pecking toys or nested dolls, such as *Chugtī Chidīyā*, *Hilāntā*, and *Sukhī Parivār*, are reliable markers of a successful process of assimilation. The new terms in the borrowing culture—in this case, *Bāṇārasi*—indicate that the borrowed products are so well accepted in the *Bāṇārasi* community that the community has severed ties with the language of the culture from which it borrowed the designs. In time, the original anglicised words will be forgotten, and the designs will become an integral part of the generational exemplars.

Heuristic 3: Assimilation is achieved better via local categories. George Lakoff (2008), in his seminal book *Women, Fire and Dangerous Things*, emphasizes that we see the world through categories. Categories establish a social structure through which a community engages with materials and people within a particular context. This process of categorization is how a community creates and sustains generational habits and sets rules for the community and its environment. Therefore, it is crucial to identify and understand these categories by examining how they persist or fade over time, the conditions under which they operate, and the effects they produce. By doing so, we can gain a comprehensive understanding of the community's practices and the dynamics that shape the community's cultural landscape.

For an outsider, such as a designer or developmental professional with a different, if not conflicting, worldview from the community's worldview, categories are not objective labels of differentiation alone. The designer's ability to use community-held categories to define and stretch the boundaries of thought and behavior establishes circles of expertise, designates areas of responsibility, and decides the measures of acceptable change. Categories are a reliable entry point into the practices of a cultural community and facilitate adoption of changes that are acceptable, appropriate, and affordable. I call them reliable because categories are set deep in the thought processes of the community and clearly establish the typical from the atypical instances of production. Positioning the intervention between a range of cultural and disruptive solutions is a call that depends on the preference for acceptability and continuity over innovation and single-use novelty. Accessible design tools that require lower learning trajectories for understanding the cultural worldview through community-held categories of objects and processes are discussed below.

(a) *Recording visual categories*: Cultural local markets are not just a point of sale but a narrative space where the stories of cultural artisanship and cultural identity come to life. The stalls are a riot of colours, bustling with eager customers bargaining over prices, a veritable din of producers crying out to woo the customers, and a localized example of merchandise display. A simple ethnographic observational record of these riotous events can provide a rich record of the culturally distinct ways in which social interactions are performed; the power structures that are maintained in transactions; and the agency of artefacts in human decision making—in this case, the shopkeepers and how they arrange and display their wares.

Often, in the cultural design communities, stalls and makeshift kiosks are set up along bustling streets or open fields, with products laid out on wooden planks, woven mats, or directly on the ground. All stalls resemble each other without window display, lighting design, or conspicuous signages dominant in the modern retail business. The typical designs are prominent in all stores in size and shape, with minor differences in colour, and are sold at the same or similar price. Most designs are historically or traditionally produced in the area, while others are slightly modified to attract the tourist population. With increasing competition, modern retail houses are striving to make their stores attractive to their target market; but here, everything is the same. The differentiating factor is not the brand the cultural community shares with each other—*Bāṇārasi*, *Morādābādī*, *Kaśmīrī*, and so on—but the familiarity with the shop owner. Patrons form bonds over time with the shop owners, sometimes between generations, and they seek goods from the same shop every year. This is a matter of loyalty and trust. These patrons do not engage in impulse buying. Their purchase is seasonal or in the context of some religious event or function.

The types of products are grouped spatially. Generally, no single product or category is displayed in a manner that would catch customers' attention first. All products are given an equal visual weightage. However, the grouping of the goods strongly corresponds with the community-held notions of categorization. Therefore, in *Bāṇāras*, *Khilōnā* related to rituals and festivals, such as *Janmashtami*, are placed differently from *Khilōnās* designed for play, such as *Chugtī Chidīyā*. Specifically, in *Bāṇāras*, the larger shops also have a room with restricted access where products designed for foreign tourists or urban clients are kept. The local patrons are not exposed to these exclusive items.

Such segregation of designs reveals the worldview of the community through its sense of grouping. It allows us to understand what is accessible or restricted, what is purchased during each season, who purchases it, weightage given to different wares, frequency of purchase, price fluctuations, and preferences within the category. Local markets of cultural design are veritable sources of visual records of community-held categories; these records are elicited nonverbally through display, consumption preferences, and practices of access.

(b) *Artefact sorting*: Similar to card sorting, artefact sorting is a tool for grouping not tasks but artefacts. The term *card sorting* applies to a wide variety of activities involving the grouping and/or naming of objects or concepts. These may be represented on physical cards, virtual cards on computer screens, or photos in either physical or digital form. For the purpose of interaction design, the sorting process—usually performed by potential users of an interactive solution—provides the following: terminology (what people call things), relationships (proximity, similarity), and categories (groups and their names). This information can be used to decide which items should be grouped together in displays, how menu contents should be organized and labelled and, perhaps most fundamentally, what words should be used to describe the objects of the users' attention (Hudson 2013).

On the other hand, artefact sorting can be employed to sort or group objects, especially in user populations that are reticent to articulate verbally or when the designer does not speak the user's language. Sorting images or artefacts does not necessarily require verbal elicitation, though its presence may further strengthen the understanding of a grouping decision. Individuals can carry out sorting independently, and their sorting decisions can easily override the researcher's influence in the study. Sorting facilitates multiple attempts of selection and decision making while using minimal resources and allowing a time gap between iterations. The iterative attempts, confusion, and time lag between decisions can give cues to difficulty, familiarity, or confusion in making a decision. The results can be expressed in several ways, with the primary focus being on which artefacts were most frequently grouped together by participants and the names given to the resulting categories.

(c) *Occupational specializations and distribution*: Within the cultural design system in India, occupational segregation is strongly practised and advocated spatially. Occupational segregation does not necessarily imply a caste system but a division of labour and practices. For instance, in *Bāṇāras*, toy designers have four kinds of skill sets: carving, turning, *pattern* making, and painting. The owners of all four skill sets live in different areas of the town; that is, they are not only temporally segregated in terms of the division of labour required for different tasks in the process of making a toy but also physically separated by boundaries of *kārkhānās* and spatial distances between each other. The turning designers live in Khojwa, while *pattern* makers can be found in Laksha, and so on.

Similarly, Mohsini (in Wilkinson-Weber and DeNicola 2020) discusses the *Zardōzī* work practiced at Seelampur in Delhi. Cultural designers often direct clients seeking cheap labour to Seelampur. Within the community, the work practiced in Seelampur is considered low in quality or, in the local language, *chalu*. For a high-quality workpiece, the order must be outsourced to senior expert designers called *Karigars* or *Buzurgs* who reside in Agra or Farrukhabad.

This spatial segregation of skill sets encourages meaningful specializations of tasks; decentralization of resources, energy, and decision making; and emulation and dissemination of novel processes or patterns. More pertinent to our discussion of categorization is the ease with which different skill sets indexing a variety of categories of objects can be identified through spatial segregation and proximity. Even a simple demographic survey of addresses of different designers mapped according to their skill sets can assist a designer in understanding the community-held categorization.

(d) *Language analysis*: The worldview of an individual is shaped and represented in language. Humboldt asserts that the mind processes the world through language. A language not only shapes the worldview, but also brings the world into being (as cited by Underhill 2009, 99). It is argued that native language not only affects our thinking in the presence or absence of a particular vocabulary, gestures, or grammatical category, but also in the manner in which we perceive events in nonverbal event representations, such as pictorial sequences. In other words, language can potentially affect our mental imagery and its signifier in semiotic media like gestures and pictures. Linguistic signifiers evolving from the local dialect influence the categories of artefacts and processes with which designers think and act. It is essential for modern designers and development professionals attempting to collaborate with cultural communities to adjust to such categorization if they are to be successful in their interactions and dealings with community members. Artefacts are shaped by a vocabulary and the worldview, but so are these shaped by the artefacts.

Drawing out a local glossary of terms helps to build a rich source of vernacular vocabularies that help us understand the notion of a proper toy for a community. Moreover, the etymological source of the vocabularies can also assist in mapping the meanings that were retained over time in the lexicon while others may have assimilated or become redundant. In examining the *Bāṇārasi Khilōnās*, we observe a significant narrowing of semantic meaning over time that transitions from divine play to a more playful, ludic interpretation. This shift has not only transformed the nature of play itself but has also redefined the players involved. The local term *Khilōnā* has evolved from a rich cultural representation into a mere ludic product and, more recently, into an item for exhibition. Such etymological shifts invite reflection on meanings that have faded as well as on the physical manifestations that may have persisted or disappeared. Notably, the category of adult play, once encompassing twenty-two distinct meanings, has largely vanished from the *Bāṇārasi* worldview. The absence of these words, which once denoted various forms of play, underscores a corresponding lack of tangible *Khilōnās* or rituals that once expressed these ideas and indicates a profound transformation in cultural practice and understanding.

Linguistic analysis holds particular significance in the context of Indian cultural design communities. While language is a common tool for elicitation in many fields of study, it has rarely been employed effectively among cultural Indian designers, primarily due to their difficulty in articulating design decisions verbally. This study introduces a method for analyzing the lexicon related to artefacts from vernacular languages and demonstrates that linguistic analysis can indeed reflect tacit knowledge and provide valuable insights into cultural design practices.

In the *Bāṇārasi* context, examining the lexicon reveals important nuances through nouns that denote artefacts, processes, and meanings. Tracking the timeline of various terms used to describe the same artefact or function unveils patterns of change in both conceptualization and physical manifestation. For instance, the evolution from *Kṛidānakā, Khilōnā to* Toy in the *Bāṇārasi* vocabulary illustrates significant shifts in cultural perception and material culture.

In the absence of explicit forms of eliciting cultural knowledge, the local *Bāṇārasi* lexicon used within the *Kārkhānā* becomes an invaluable source of insights. These linguistic cues help capture the worldview of *Bāṇārasi* toy designers and offer a deeper understanding of their design philosophy and practices. Thus, the success of this method challenges the perceived limitations of language in conveying tacit knowledge and opens new avenues for studying cultural design traditions.

Heuristic 4: It is helpful to evaluate by examining emulation. From the point of view of the designers and developmental professionals striving to engage with cultural design practices, the act of emulation emerges as a profoundly insightful communal endeavor. Emulation is a practice of learning as well as upgrading cultural knowledge in the design communities. The outcomes of emulation, however, are contingent on who is emulating whom. Apprentices follow in the footsteps of their master as closely as possible to gain membership in the community; they adopt the community's shared conventions, canons, and beliefs in their skill set. Gaining membership is not a matter of mastering a skill alone but of imbibing the worldview and the ability to practice it. However, apprentices are not the sole practitioners of emulation within the community. Masters, too, engage in emulation, often modelling their actions after other masters. Masters emulate other masters to gain from their experience in how to evaluate apprentices and to correct, rectify, or applaud their attempts to scaffold their learning trajectories. A seasoned master might emulate another to assimilate or refine knowledge of innovative designs or processes that their peers have encountered. Occasionally, a master is presented with the fortuitous challenge of a substantial client order. In the absence of mass production tools and machinery, such an order cannot be fulfilled by a single workshop alone. In these instances, the master may solicit other masters to emulate the client order as closely as possible, thereby ensuring that the production demand is met within the stipulated deadlines. Through these intricate layers of emulation, the practice not only fosters individual growth but also fortifies the communal bonds and collective expertise that sustain cultural design practices.

Understanding the dynamics of who is emulating whom and the specific purposes behind these emulations can provide modern professionals with profound insights into the community's perception of what is typical and significant. A crucial indicator of the acceptability of a participative outcome. The new statement will read as- A crucial indicator of the acceptability of a participative outcome in cultural communities is the community's decision to emulate the outcome. cultural communities is the community's decision to emulate the outcome.

When community members begin to emulate a product or the process of selling it, whether within or beyond their markets, they signify their voluntary acceptance of the changes. This emulation suggests that they will continue to adopt these changes even after the external agency has withdrawn. If the masters within the community integrate the new changes into the pedagogical practices of the *kārkhānā* while training apprentices, then the changes endorsed by the participative outcome will transcend generational boundaries.

I strongly advocate for incorporating the emulation of changes into the impact evaluation of all developmental interventions in cultural design communities. This practice will help assess the sustainability of the participative project. A product or process that is not emulated by cultural designers will die naturally after the external agencies withdraw themselves from the community.

Object geographies and itineraries are vital in understanding cultural design processes. Cultural designs stand out for their remarkable continuities across geographies, chronologies, and cultures. Mapping the journeys of objects from their creation through the incremental, assimilative, and disruptive changes they undergo as they are handed down between generations or cultures reveals intricate networks of trade, exchange, and cultural interaction that mold cultural societies. Tracing the mobility of these objects offers external agencies a powerful tool to comprehend how worldviews are formed and to what extent objects shape and are shaped by these worldviews in return.

By examining these trajectories, we can uncover the layers of historical and cultural significance embedded in cultural designs. Each object, as it moves through time and space, accrues stories and meanings that reflect the diverse influences and exchanges it has encountered. This approach not only highlights the resilience and adaptability of cultural designs but also underscores their role as active participants in cultural dialogue.

Understanding these patterns of mobility and change provides modern professionals with valuable insights into the underlying principles and values that sustain cultural communities. It allows for a deeper appreciation of the ways in which these designs are not static relics of the past but dynamic entities that continue to evolve and influence contemporary practices.

Participative initiatives could aim at revealing the narratives stored within the objects to gain community-driven insights into the diverse paths in which community members think, act, and interact with objects. Cultural design communities have an affinity to continuities, and locating and identifying design trajectories of objects is a matter of foraging through the sample sets visible in the *Kārkhānās*, local shops, and storehouses. These continuities are also available as exemplars in museum displays and in the collections of aficionados. In India, old exhibition catalogues such as that for the India Exhibition of 1886, gazetteers, and the *Shilpi* award list of the Government of India are also probable sources of continuities. Oral records of the cultural designers can often accurately recollect the continuity of at least the past generation, if not older ones.

In this study, I have reconstructed the itinerary of the pecking toy. I collected samples of these toys via direct purchase from the *Bāṇārasi* shops, from antiques warehouses, or from toy designers themselves; I also gathered images of toys from the archives of the Banaras Hindu University, the Warwickshire Museum Service, the Victoria & Albert Museum (V&A), and the Craft Museum in New Delhi (all toys were created in the period from 1900 to 2011). Despite acquiring several instances of the pecking toy from various periods, the origin of the pecking toy in *Bāṇāras* was difficult to determine; however, I acquired a sample from an expert order from a client, and via this sample I was able to determine that the pecking toy had been assimilated in the *Bāṇārasi* toy repertoire during the early 1920s. Over time, pecking toys and their variations have acquired indigenous names such as *Chugtī Chidīyā*, which is a literal translation of the English term *pecking bird*. Other variations are called *Hilānā*, which is drawn from the verb *Hilāntā* and means "to move." Both the tail and the head follow an up-down, vertical trajectory. The toy is often placed on a wooden base. If this base is triangular, it has three birds; if octagonal, it has eight birds; or if the base is a single rectangle, it has one bird or a pair of birds (refer to Fig. 2, Patil 2017).

Another compelling example of how object itineraries illuminate a community's design trajectories is the story of the *Sukhī Parivār*, an indigenous version of the Russian nested doll. In the 1930s, one of the early toys borrowed from *Bāṇārasi* designers was a client order for Russian nested dolls. Initially, the design was replicated to meet the client's specifications. However, with sustained exposure, elements of the toy were assimilated into a *Bāṇārasi* version, which culminated in the *Sukhī Parivār* (Happy Family) in the 1980s.

Sukhī Parivār consists of a father figure within whom a mother and two children are nested. (Refer to Fig. 3.) This concept originated from the Indian government's

Fig. 2 Object itinerary of the pecking toy in *Bāṇāras*. *Image source* Author

family planning campaign, *Hum do hamāre do* ("We two and our two"). The *Bāṇārasi* version of the Russian nested doll retained the nesting mechanism of the Russian doll, the material, and its manufacturing process. But the *Bāṇārasi* toy designers added their own meaning through *Bāṇārasi* characters and themes that were more appropriate to the community worldview. By adopting the theme of the government campaign on family planning, the toy designers nested different characters—father, mother, and two children—within the largest doll. This is unlike the conventional Russian dolls, which often use the smaller version of the main doll to nest within. More significantly, unlike the client-ordered Russian dolls, the *Sukhī Parivār* has become a quintessential *Bāṇārasi* doll (Patil and Athavankar 2022). These dolls are produced and sold for the local market as well as for export. The object itinerary scaffolds the design understanding of designers and developmental professionals regarding the features that form the unique traits of the *Bāṇārasi* design. This understanding has led to the essential design being retained despite iterations, the extent of change or modification acceptable in the community, the pace of change, geographical influences, and chronological interpretations and representations. Hence, object geographies and itineraries are constructive tools for modern designers in decoding the nature of acceptable change in cultural design communities.

Heuristic 5: Recording the typical helps in identifying the acceptable: The quintessential marker of acceptability in cultural design is the typical member of any category. As discussed in Chap. 5, these typical members serve as prototypical examples of their category based on their perceived or actual benefits or value to the community. In her seminal book on prototype classification, Rosch (1973) advocates that certain stimuli are better examples of a concept than others. According

(a) (b) (c)

Fig. 3 **a** Mūrti and nested dolls made in *Banāras* since the 1920s. **b** *Hum do hamāre do* postal stamp. **c** *Sukhī Parivār* nested doll. *Image source* Author; also published by the author in Patil and Athavankar (2022)

to the prototype theory, a prototype is the central tendency of a category and represents the best or most typical example of the concept. This prototype (a) shares the maximum number of features or attributes with the category, and (b) has a minimum number of instances of other concepts. Understanding these prototypical members helps designers and developmental professionals identify the core characteristics that are most valued within a community. This knowledge is crucial for creating designs that resonate deeply with the community and ensuring that innovations are aligned with cultural values and practices. Modern designers and development agencies can discern cues of the typical members as follows:

(a) *Frequency and seasonality of production*: The spinning tops or the pecking toys are made more often than most other toys in *Bāṇāras* and are among the oldest designs still in production. These toys are very popular among consumers and are sold throughout the year. Unlike seasonal toys, these prototypical instances are always in demand. The frequent exposure of these toys makes them prototypical instances of the *Bāṇārasi* toys. Their continual production and demand underscore their role as central markers of *Bāṇārasi* toy-making heritage, and as such they reflect both the community's values and the consumers' preferences.

(b) *Part of the learning pedagogy*: Master toy designers opt to teach the skill of toy making through the most common and typical instances of toys familiar to the community. These prototypical toys are often small in size and feature large, plain surfaces that do not require highly refined and controlled motor skills. Their quick turnaround time further makes them ideal for teaching purposes. The production process of these toys has been honed through trial and error, which makes them robust and adaptable to various contingencies. Consequently, there are minimal risks of failure, and any failure that does occur does not significantly impact the economic well-being of *Kārkhānā* because the process is inexpensive.

    These prototypical toys are accessible to everyone within the learning community and thus offer a practical and efficient means for apprentices to develop their skills. They are reusable and interoperable across all studios, masters, and learners, which ensures a consistent and collaborative learning environment. The choice of these toys for teaching reflects their essential role in preserving and perpetuating cultural design because they embody the core principles and techniques that define *Bāṇārasi* toy making.

(c) *Ease of recall*: Typical instances are distinctly recognized as representative of the wooden toy maker's cultural design. Many consumers knowledgeable about cultural design traditions can easily identify these prototypical examples as part of *Bāṇārasi* heritage. The ease with which some designs are recalled over others further underscores their typicality. This recall can be examined through oral narratives , and pictorial catalogues. Oral narratives, rich with numerous instances of objects that community members have interacted with during their childhood or adulthood, play a crucial role in this identification process. These stories not only trace the itineraries of objects but also highlight the most memorable instances. Through these narratives, the community's engagement with these toys is vividly remembered and passed down, thereby reinforcing the cultural significance of these prototypical designs.

Pictorial catalogues also serve as a valuable resource because they offer visual documentation that complements the oral narratives. Together, the narratives and catalogues provide a comprehensive understanding of which designs have stood the test of time and why they hold a prominent place in the collective memory.

Conventionally, the design worldview has adopted a dualistic position between the worldview as a neutral, external background to the community activities and a deterministic view that worldviews are an unchanging and inherent cultural ordering of things, relationships, and agents. Rather, I adopt the perspective of Tim Ingold, that design worldviews are neither of these two; they are instead a "dwelling perspective," which constitutes "an enduring record of—and testimony to—the lives and works of past generations who have dwelt within it and, in so doing, have left there "something of themselves" (Ingold 1993, 153). That worldview of cultural design communities as dwelling places can include "views of guests" like me and other development professionals. Dwelling perspectives are enacted in "taskspaces," again a term coined by Tim Ingold to signify a socially constructed space that features an array of related human activities possessing spatial and functional boundaries. *Kārkhānās* are instances of taskspaces that exhibit continuities in process and are shaped by a mutable dwelling perspective. The interiors of such dwelling views can be redesigned part by part without destroying their very existence through sensitive, respectful, and responsible engagement with the original inhabitants of the dwellings.

References

Balaram S (2005) Design pedagogy in India: a perspective. Des Issues 21(4):11–22

Balaram S (1998) Thinking design. National Institute of Design, Ahmedabad

DeNicola AO (2005) Working through tradition: experiential learning and formal training as markers of class and caste in North Indian block printing. Anthropol Work Rev 26(2):12–16

Hubka T (1979) Just folks designing: vernacular designers and the generation of form. J Archit Educ 32(3):27–29

Hudson W (2013) Card sorting. In Soegaard M, Dam RD (eds) The encyclopedia of human-computer interaction, 2nd edn. The Interaction Design Foundation, Denmark

Ingold T (1993) The temporality of the landscape. World Archaeol 25(2):152–174

Lakoff G (2008) Women, fire, and dangerous things: what categories reveal about the mind. University of Chicago Press

Naji M (2020) Creativity and tradition: keeping craft alive among Moroccan carpet weavers and French organic farmers. In: Critical craft. Routledge, pp 153–167

Papanek V, Buckminster Fuller R (1972) Design for the real world

Patil K, Athavankar U (2012) Un-authored artifacts: it takes a whole community to make a Khilona. In: Participatory innovation conference. University of South Denmark Melbourne, Australia, pp 37–45

Patil K, Athavankar U (2022) Design precedents to design innovation: category-based reasoning in problem-solving. Des J 25(1):62–85

Patil K (2025). Artisanship of risk: the uncertainty of traditional design. Journal of Vocational Education & Training 1–22

Patil K (2017) Inexplicit learning: transferring knowledge through visual and emulative practice. Techne Ser—Res Sloyd Educ Craft Sci A 24(2):142–164
Pound E (1934) Make it new. Faber and Faber, UK
Rosch E(1973) Prototype theory. Cognitive development and the acquisition of language, pp 111–144
Roy T (2002) Acceptance of innovations in early twentieth-century Indian weaving. Econ Hist Rev 55(3):507–532
Stella K (1958) Traditions of the Indian craftsman. J Am Folklore 71:224–230
Stephen L (1991) Culture as a resource: four cases of self-managed indigenous craft production in Latin America. Econ Dev Cult Change 40(1):101–130
Underhill JW (2009) Humboldt, worldview and language. Edinburgh University Press, Edinburgh
Ward LM, Daniels EA, Zurbriggen EL, Rosenscruggs D (2023) The sources and consequences of sexual objectification. Nat Rev Psychol 2(8):496–513
Wenger E (1999) Communities of practice: learning, meaning, and identity. Cambridge University Press
Wilkinson-Weber CM, DeNicola AO (eds) (2020) Critical craft: technology, globalization, and capitalism. Routledge

Chapter 8
Emerging Signifiers of Cultural Design

1 Multiple Signifiers of Cultural Design

Cultural design, which is often discussed as vernacular design, craft, handmade, decorative arts, un-self-conscious design, pre-modern design, slow design, and so on, is a messy word akin to another messy word that often has a multitude of meanings: *wicked*.[1] Cultural design has become a category in opposition to that of art and design but is also an umbrella term encompassing a range of disparate activities that show no intrinsic affinity in their products, processes, skills, and ideas. The term is often applied loosely to describe practices ranging from traditional craft to modern reinterpretations of heritage, and from community-based making to globally commodified objects with a cultural signature. The lack of coherence among the practices grouped under this category indicates its fragmented nature. As Greenhalgh (2002, 1) states:

> They [crafts] have no intrinsic cohesion; they have no a priori relationship that makes them a permanently peculiar or special gathering; there could be fewer or more of them; they are together, as we move into the next century, because the complex forces that brought them together, despite shifts in circumstance, hold them in proximity. This proximity is not stable and is certainly in the process of change. Nevertheless, they are in a consortium still.

Depending on how we have historically defined cultural design, our perception of how cultural designers are represented, their status, means of evaluation, needs, and what makes them thrive in contemporary times has remained in flux. As the industry's

[1] In 1973, Rittel and Webber, in their seminal paper "Dilemmas in a General Theory of Planning," elaborated on ten characteristics of the wicked problem. They said wicked problems (1) do not have a definitive formulation; (2) do not have a "stopping rule"; (3) do not have solutions that are true or false, only good or bad; (4) do not have a way to test their solutions; (5) cannot be studied through trial and error; (6) have no end to their number of solutions or approaches; (7) are essentially unique; (8) can always be described as the symptom of other problems; (9) describe themselves to determine their possible solutions; and 10 have planners—who present solutions to these problems and who have no right to be wrong.

K. Patil, *Toying with the Mind*, Design Science and Innovation,
https://doi.org/10.1007/978-981-96-9227-9_8

"other' (Adamson 2007), the discourse on cultural design started as recently as the Industrial Revolution and was contrasted with the discourse on machines. Admired as an "artisanship of risk" (Pye 1968), discussions of cultural design practices have produced a plethora of arguments to establish that the human hand is not merely imitative but also innovative. Summed up as an un-self-conscious design, these practices are considered rigidly bound to culture and constantly improvising to meet small contextual changes. Cultural design is regarded as an "informal practice" (Reitan 2006)—a way of learning and producing outside the formal educational framework. Perched on a pedestal, the cultural designer is an instrument whose skills originate from a divine source or a fountainhead of creativity (Coomarswamy 1909), marking him as the "chosen one" or the "gifted one." From the perspective of vernacular understanding, cultural design contests the hegemony of the modern or the mainstream. It has also been dismissed as a pastoral nostalgia or an airport trinket. However, more recently, it has also been applauded as "an enduring, basic human impulse, the desire to do a job well for its own sake" (Sennett 2008, 9).

Based on these definitions, contesting conclusions can be drawn regarding cultural design, cultural designers, and their artisanship. One can infer that cultural design and cultural designers are marginalized victims of industrialization and modernization; embodiments of an enduring and living practice of making; a reviving culture to be conserved and protected; a resilient impulse to respond to materials; an anthropological signifier; or a sustainable alternative to the modern-day environmental crisis. "There have been so many worthy agenda and maundering laments that we barely know what [cultural design] is anymore, and a lot of sensible people no longer care" (Greenhalgh 2002, 1).

Marchand acknowledges that cultural design is a polythetic category, "one in which any of its members possess some, but not necessarily all, the properties attributed to that category" (2016, 8). Understanding this discipline as an exclusive category but also a disparate range of practices and positions has raised contestations about what cultural design represents, its status, and the reasons for its resilience (Greenhalgh 2002, 16). In this text, I do not wish to debate the definitions of cultural design (several scholars have written extensively on this subject). Several signifiers of this category have been elaborated in each definition, and these signifiers are necessary but not sufficient. The latter has been a challenge given the very wide range of activities that fall within the purview of this category—from a conservative set of pottery, basket weaving, or carpentry to the contentious inclusion of horse training (Crowder 2016), surgery, and parenting (Sennett 2008, 101–102) as cultural design.

Based on my close study of the *Banārasī* wooden toy designers, I would like to enhance the set of characteristics by which we define, examine, intervene, and adjudge the future viability of cultural design. As Marchand (2016, 10) stated:

> Yet further attributes may be included and the polythetic nature of craft licenses this. The list is by no means exhaustive, and its contents are liable to further expansion (or contraction) in response to changing technologies for making, introductions of new materials, and changing political landscapes, economies, and regimes of consumer taste at both local and global levels. All such changes throw up fresh challenges to overcome, new kinds of mistakes to

> be made and repaired, and novel sorts of problems to solve. Craft meaning, identity, and knowledge are in a state of constant evolution.[2]

This chapter attempts to increase the number of signifiers and thereby to make this polythetic category more inclusive, semantically rich, and possibly more holistic.

2 Excellence of the Typical

If you ever visit the Nara temples in Japan, look for not just the Temples, the deer, and the manhole covers, but also for a quaint little shop called *Nakatanidō* that is just outside the gates of the Kōfukuji Temple. This unapologetically spartan shop is famous for a delicacy called *Yomogi mochi*. Produced with Japanese mugwort, these verdure rice cakes are filled with a delicious crimson paste of azuki beans and drizzled with roasted, sweetened soybean flour (*Kinako*). But the secret of this chewy, aromatic delicacy lies not only in the ingredients but also in the act of pounding it.

The age-old *Mochi Tsuki* pounding technique transforms the sticky, glutinous rice into a smooth, springy paste that is used to cook the rice cakes. The mixture is first battered with *kine* or wooden mallets. Then, an expert pounder (called *Kaeshi-te*), in this case *Nakatani San*, manually stirs the paste between the blows as another worker (called *Tsuki-te*) continues to pound the mallet at an increasing speed. The rhythmic process requires an intimate understanding of the material, the action, and the body of the partner. The skilled performers possess such proficiency that they can quickly and accurately pound *mochi* without ever skipping a beat, and they let out a yell with each swing of the heavy pestle. The speed of this pounding performance can be an astonishing three times per second.

Every year, several *Mochi Tsuki* pounding competitions are held all over Japan. Each pair of *Tsuki-te* and *Kaeshi-te* pounds and stirs the paste manually, speeding up their performance to compete with other pairs. They are all competing to be better at the same task, using the same technology, and with the same skill set. The artisanship does not lie in making a different *Mochi* but in making the same *Mochi*, with the same ingredients, process, and technology, albeit better than all other chefs in the community. All the chefs are striving to achieve the distinction of becoming the typical *Mochi Tsuki* pounders who make the typical *Mochi*. Currently, this distinction belongs to *Nakatani San* and the *Yomogi Mochi*.

The bustling old city of Lucknow, India, much like the vibrant *Yomogi Mochi* stalls, is lined with numerous small and large shops, each asserting its claim as the original, time-honored establishment of *Ṭunday ke Kabāb*, which purportedly dates back to 1905. This Awadhi delicacy, *Ṭunday Kabāb*, is crafted from minced meat and is renowned for its intricate blend of over a hundred spices. Yet locating the true original shop that opened in 1905 in the labyrinth of Lucknow's old city is a

---

[2] Marchand (2016) notes this in response to a conservative alphabetical inventory of things, properties, and characteristics attributed to cultural design as listed in the introductory chapter of his book and posits craftwork as a problem-solving activity.

nearly impossible feat. Each shop professes to uphold the same legend and aroma and creates a nearly universal imitation of the original. The relentless pursuit to replicate the oldest recipe drives each vendor to follow what is likely the same formula and to strive to re-create the dish in its purest form. In this context, no shop seeks to be distinct; rather, survival hinges on each shop's ability to mirror the supposed original.

This phenomenon in cultural design is particularly intriguing because excellence is attributed not to deviation but to the finest iteration of the typical. This notion stands in stark contrast to modern design and art, where distinction is often achieved through differentiation. In cultural design, however, the typical is synonymous with excellence. The typical is not a mundane, disposable product but rather an earthly *eidos*—a type that embodies a meaning that the community regards as essential or remarkable. Consider the *Yomogi mochi*, *Ṭunday Kabāb*, *Banārasī* wooden *Jhanki* toys, muslin *Jāmdānī* sarees (Chauhan 2022), *Kizaemon* tea bowl (Yanagi and Leach 1989), *Iñupiaq-Inuit* clothing (Reitan 2006), *Matryoshka* dolls (Ertl and Hibberd 2003), among countless others you could surely add. These are all archetypal examples within their categories, and community members are committed to achieving the standard set by such objects. In cultural design, the typical represents a standard that has been perpetuated across generations, and it is this standard of excellence that all participants strive to practice, replicate, and attain. Emulating the typical member is not only unashamedly accepted but is also free from legal repercussions and, therefore, serves as an open source of expertise.

However, it becomes evident that mere cognitive comprehension is insufficient for the creation of an artefact. Cycles of approximation, continual prototyping, the 10,000-hour rule of practice, practice by repetition, and emulation underscore the necessity for embodied knowledge—the transfer of explicit information into the tacit understanding of the practitioner. Thus, achieving the typical is not a matter of mindless replication but an explicit and deliberate objective that requires a certain sensory dexterity paired with a mindful understanding of the past. It is an individual's intrinsic drive, not external pressures, that compels them to continually refine and enhance the typical product until the limits of the human hand, mind, and tool are fully exhausted. Only at this point is a stopping rule applied to the design; nonetheless, other community members persist in testing their limits and strive to surpass the benchmarks set by their peers or predecessors.

This perspective invites us to reconsider our romanticized image of the genius or expert as a solitary, eccentric maverick—a notion rooted in the disciplines of art, science, and design, but often uncritically adopted in cultural design. A revised understanding of a master in cultural design might instead encompass a mindful individual as well as a community, both of which are distinguished by their ability to finely tune and discipline their senses. This allows them to engage intuitively with materials, tools, and techniques to achieve the most ingenious and typical resolutions to the normative conflicts of daily life and the present moment. These resolutions, in turn, are embraced by the entire community as authentic and exemplary and as representing the finest-crafted instance of their kind.

3 The Stopping Rule

Evolutionarily fit forms do exist, though not within the realm of modern design, where the relentless drive for improvization and novelty pushes the industry to continually innovate or create parity products. These modern designs often tackle wicked problems—complex issues with no clear stopping rule, as articulated by Rittel and Webber (1973). Such problems in modern design lack a built-in logic or argument that indicates when they are resolved. In contrast, cultural design, as explained by Alexander Christopher (1964), emerges from a process where rigid traditions and constant improvizations to meet changes in the environment, lead to a "good fit" (28–46) with the context. Once this good fit is achieved, the evolved design remains unchanged for generations due to the slow pace of change within the community. Preserving these designs becomes a cognitively economical strategy, akin to the principle of survival of the fittest. What has endured in nature has often been conserved for centuries.

A striking example of a stopping rule in cultural design is the *Kulhad*—an undecorated, unglazed terracotta cup used for drinking tea. The earliest conical, handle-less versions of these cups date back to 3300 BCE in the Indus Valley civilization. Their production involves the simplest hand movements on a potter's wheel and uses locally sourced materials and basic technology that emerged sporadically worldwide since the Jomon period. The *Kulhad* is a disposable, biodegradable cup that fits comfortably between the palms and provides both refreshment and warmth. As the techniques for making it spread, production became decentralized, and the design innovations ceased. Despite changes in technology, materials, and consumption patterns, this evolutionarily fit design has remained unchanged for centuries because it adhered to a stopping rule.

Consider another instance. In April 2020, the Crafts Study Center and University for the Creative Arts, Farnham, United Kingdom, invited three contemporary makers to draw inspiration from a wheel dating back to 1894 from their museum collections. The question arises: Can one truly improve, let alone invent, a new wheel? Gareth Neal, one of the designer-makers invited for the exhibition, pondered, "It [the wheel] is the result of an accumulation of knowledge and experience over centuries by wheelwrights who spent their whole lives mastering the craft." The four components of the wheel—hub, spokes, felloes, and rim—have been iterated through generations of wheelwrights and have reached their evolutionary saturation. Only incremental changes, if any, are now made, and the design has reached a point where further innovation has ceased.

It can be argued that cultural designs are imitative and repetitive; however, alternatively, these practices might be seen as striving for an elusive perfection achievable only through a singular focus on one object, not just by an individual but across generations. In 1993, Anders Ericsson, Ralf Krampe, and Clemens Tesch-Römer's work, *The Role of Deliberate Practice in the Acquisition of Expert Performance*, introduced the concept of the 10,000-h rule. The study discussed in the paper involved a group of psychologists in Berlin who conducted a longitudinal study of violin students from childhood through adulthood. Initially, children begin playing at around age five, have similar practice hours, and repeat the same notes as prescribed in violin classes across generations. However, by age eight, their practice hours begin to diverge. By age twenty, gifted performers average around 10,000 h of practice, while less proficient performers average around 4000 h. Ericsson concluded that "many characteristics once believed to reflect innate talent are actually the result of intense practice extended for a minimum of 10 years." Imagine these 10,000 h repeated over generations—such repetition might yield that elusive perfect symphony. Once the music community reached *Symphony No. 5* (also known as the Fate Symphony composed by Ludwig van Beethoven), there was no further improvization; they simply repeated it. They adhered to a stopping rule.

4 Explicit-to-Tacit Practice

Since the work of Donald Schön (1983) and Nonaka and Takeuchi (1995), there has been a pronounced focus on converting tacit knowledge into explicit knowledge that can be used for transferring information and skills from one individual to another. With frequent employee turnover, companies not only lose human resources but also "the tacit and often highly subjective insights, intuitions, and hunches of individual employees" and the ability to make "those insights available for testing and use by the company as a whole" (Nonaka and Takeuchi 1995, 9). The Western approach, from Henry Ford to Steve Jobs, has been to formalize, structure, rigidly categorize, set hierarchies and disciplines, codify procedures, and convert numerical data into decisions, thereby aligning with the analogy of an organization as a data processor. Thus, making all knowledge, information, and decisions explicit and recorded for further action is considered crucial. Such a machine-like organization struggles to accommodate human heuristics, intuitions, metaphors, and inductive thinking—an insight Hubert Dreyfus and Stuart Dreyfus had in 1986, when they argued that computers are inherently inferior to the human brain. In this machine analogy, the focus is on making tacit knowledge explicit, and measures for evaluating mastery are well-defined and quantifiable—all with the goal of improving efficiency, lowering costs, and increasing profits.

However, in cultural design, mastery lies not in explicating but in enhancing one's repertoire of tacit knowledge and skills. This limitation was evident in the inability to conduct a protocol analysis on the *Banārasī* toymakers to trace their thought patterns, a challenge similarly applicable to modern designers. In their seminal work on artificial intelligence, Herbert Dreyfus and Stuart Dreyfus (1986, 31–36) argued that computers cannot rival the human mind. While computers are efficient at generating and reading patterns, expert humans rely on intuition—an ability that defies explanation. Expertise manifests in "fluid performance." Experts achieve "experience-based holistic recognition of similarity" that leads to deep situational understanding. Expert performance is generally "ongoing and non-reflective," but in critical or emergency situations, experts deliberate and "critically reflect on one's intuitions" before acting.

For example, an expert typist types effortlessly without seeing the keys due to familiarity and muscle memory, and performs novel key combinations without conscious judgment. Typing, like walking, cycling, or swimming, becomes a tacit skill—muscle memory. Interestingly, akin to Christopher Alexander's un-self-conscious designer (1964), neuroscience studies also suggest that muscle memory is stored unconsciously; skilled actions can be performed automatically without conscious awareness.

In a Patil anbd Uday (2019) paper, the author and Professor Uday Athavankar discussed a small protocol study conducted with an expert traditional potter. The study involved two experiments with stimuli of varying complexity. In both experiments, the potter successfully created a clay pot on the wheel from a given 2D drawing stimulus. The subject was then blindfolded and asked to recall and trace the procedural sequence of the two clay pots gesturally, without the use of clay. Remarkably, the potter hand-held the spatial information and created a virtual pot in the air. Even without tactile and visual feedback, the expert was able to reconstruct the transformation of the 2D stimulus from a lump-like form into a finished pot by accurately recalling both procedural and spatial information. Even more surprising was the potter's ability to evaluate and correct errors in this virtual pot while blindfolded. This study sheds light on the power of muscle memory and tacit problem solving. The hand skills are so deeply embedded in the long-term memory of cultural designers that they can recall and manipulate them even under unusual conditions, such as being blindfolded.

In such a state of flow, there is no distinction between thinking and doing because thought itself requires time and effort. My Tai-chi teacher would often have me repeat the form with the instruction that I should be able to perform it flawlessly, even if awakened at 2:00 a.m. He insisted that the form should become ingrained in my muscle memory, which would allow its performance to be immediate and without the delay of conscious thought. Dreyfus and Dreyfus (1986) and my Tai-chi teacher advocate the same principle: Practice and repeat until the explicitly taught instructions have been internalized as tacit knowledge. The golden 10,000-h rule for achieving mastery is essentially the number of hours required to develop an intuitive understanding of practice. In body-dominated practices, mastery is derived from tacit knowledge, and the inability to make such knowledge explicit is neither a limitation nor a flaw.

5 Precedent-Based Problem Solvers

In his post-Promethean theory of action, Bruno Latour emphasizes that there is always something remedial in design (2008, 4). Design is never a process that begins from scratch: to design is to redesign because there is always something a prioria—an alternative, issue, or problem. The bulk of human problem solving occurs in problem spaces that are either well known or vary only slightly from familiar situations. Thus, it is rare to encounter a problem that bears no relation to similar problems attempted or resolved in experience and that is not stored and retrieved when required from the design memory (Carbonell 1982; Doboli & Umbarkar 2014; Eilouti 2009). Yet modern design is driven by the overwhelming desire to create something new that has never existed. Severing ties with the past had been a key attribute of the modern movement in the immediate wake of the Industrial Revolution. The "moderns" have an "anxiety of influence" (Bloom 1997). Bloom explains the anxious struggle of newer poems to clear creative space for themselves from the older ones. There is a fear of being engulfed, annihilated, and overwhelmed by the very giants on whose shoulders we sit. That is why the creative space must be "cleaned" of its predecessors.

Cultural design, on the other hand, acknowledges and adheres to the precedents. It shows no anxiety about its past. It is a strategy to maintain the product's identity and artisanship, a method of incremental and sometimes disruptive innovation, a means to preserve knowledge without any explicit means of maintaining records and transferring know-how between generations. It is a generational practice of iterations to evolve a form that is a good fit to the context. The sometimes-evolutionary nature of the good fit also supports the exercise of a stopping rule.

David Gates (2016) is a designer-maker who seamlessly integrates both modern design techniques, such as pre-planning and sketching, with practices rooted in cultural design, like iterative prototyping. He refers to this approach as "perpetual prototyping" and emphasises the ongoing nature of refining and improving his work. Much like George Sturt (1963), Gates provides an anthropological account of his practice as embodying the role of a conscious cultural design maker.

> Not all problems that a cabinet maker will confront in making the piece can be foreseen at the design stage. Intentions can be stated with some surety on the rod drawing, but the more complicated matter of "how" something will be made and assembled may only become apparent once enmeshed in the making. To some extent, decisions of "how" can be mitigated by the trust that an experienced cabinetmaker places in their previous experience and existing ability. No piece of furniture is so entirely different from the pieces that they will have made in the past. Prior experience will therefore inform their approach to a present problem, at least in some way. (Gates 2016, 118)

In 2011, Ratanlal Prajapati—a toy designer from Banāras—was asked to redesign the famous *Banārasī flying toy*. (For a more detailed analysis of this case study, please refer to Patil and Athavankar 2017.) *Banārasī flying toys* are immobile figurines carved in solid wood, hung from the walls, and consisting of deities riding their *Vāhan*[3] (see Fig. 1a) Ratanlal Prajāpati was commissioned to innovate a novel mechanism that would move more than two body parts of the *Vāhan* simultaneously. Simultaneous movement on multiple axes is difficult because the proposition requires additional hinges and coordination between threads to synchronize the movement of all parts. Remarkably, within just ten days, Ratanlal integrated the traditionally well-known and practiced pendulum and hinge mechanism of the *Hilāntā* toy with the static flying toy, which resulted in an innovative solution. In a notable departure

[3] A *Vāhan* is an animal designated to each deity in Hindu mythology for travelling on land, water, and sky. For instance, Viṣṇu glides on a *Garuḍa* or an eagle, Ganeśa skitters on a *Māśuka* or a mouse, Kārtikeya flies on a peacock, Sūrya gallops on a horse. Any change between these pairs of deity-*Vāhan* is perceived as a sacrilege in the *Banārasī* community.

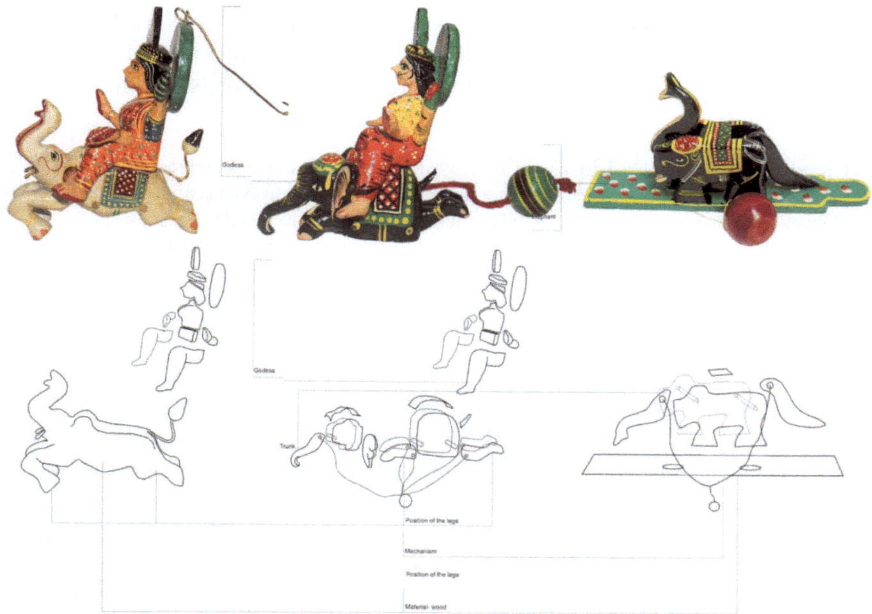

(a) Precedent source Decoration *kā khilōnā*

(b) Targeted source

(c) Precedent source *Khelne kā khilōnā*

Fig. 1 Precedent-based innovation of the new *flying toy* with moving parts. *Image source* Author: also, published by the author in Patil and Athavankar (2017)

from the *Hilāntā* mechanism, Ratanlal aligned the movement of the head and tail of the bird (*Vāhan*) on the same axis and perched the wings on an axis perpendicular to it. This simultaneous movement on multiple axes necessitated the addition of extra hinges and the careful coordination of threads to synchronize the up-and-down movement of all parts. (see Fig. 1 and refer to Patil 2017). Ratanlal's case highlights the extensive repertoire of templates that cultural designers possess in their minds and their ability to retrieve and adapt them to address contemporary problems or tasks. Furthermore, contrary to the common belief that precedent-based learning results in only incremental changes, Ratanlal's innovation demonstrates that novel combinations of precedents can lead to moderately disruptive solutions in cultural design.

6 Emulative Practices

Well before the electronic revolution, *Shānzhài* simply signified a mountain fortress or a political stronghold. However, its use from the early 2000s onward usually encompassed counterfeit, imitation, cheap, fake products or the underground culture surrounding such products. Beginning with DVDs and MP3 players, in 2010, unknown engineers with little-known tinker shops accounted for about 20 percent of the global mobile phone market. Yu Hua records that *Shānzhài* is not limited to mobile phones:

> Once copycat cell phones had taken China by storm, copycat digital cameras, copycat MP3 players, copycat game consoles and other pirated and knockoff products came pouring forth. Copycat brands have rapidly expanded to include instant noodles, sodas, milk, medications, laundry detergent and sports shoes, and so the word "copycat" has penetrated deep into every aspect of Chinese people's lives. Copycat stars, TV programs, advertisements, pop songs, Spring Festival galas, Shenzhou space capsules and Bird's Nest national stadiums have all made a splash on the Internet, each revealing their own special flavor and gaining instant popularity. (Hua 2011, 185)

The concept of *Shānzhài* has sparked mixed reactions, with some perceiving it as a mere synonym for Chinese knock-offs and counterfeit goods and others recognizing it as a testament to the ingenuity and adaptability inherent in Chinese artisanship, which includes prowess not only to re-engineer but also to tinker and innovate upon the re-engineered technologies. *Shānzhài* is not a fake in the Western sense because in the Chinese culture (and so in Japanese culture; see Guth 2010; Madeley 2007) imitating a master's work is regarded as a valuable pedagogical tool. The ability to replicate a master's work is not only a mark of respect but also a sign of prowess and a means of earning admiration within the community.

Moreover, *Shānzhài* represents a form of resistance against the dominance of global brands, which are often unaffordable to the average Chinese consumer. These inexpensive imitations serve as a Robin Hood–like solution by providing the marginalized masses with access to products that would otherwise be out of reach. In fact, the earliest customers of *Shānzhài* products were immigrant workers from rural areas who moved to cities in search of better opportunities. These consumers found value in *Shānzhài* goods not just because they were cheaper, but because they often offered enhancements beyond the original products. For instance, the Anycool V876, a *Shānzhài* version of Samsung's Anycall, included additional features that the original did not and made it more attractive to buyers—it was less expensive, had a similar brand identity, and offered more functionality.

From the 1990s onward, as post-nationalistic sentiments grew, an increasing number of Chinese consumers began to favor *Shānzhài*—local or indigenous goods—over imported ones as a demonstration of loyalty to their nation. This shift has allowed several *Shānzhài* entities to evolve from copycat operations into legitimate businesses that now compete with global brands, often with entirely new product lines. A prime example is the company Build Your Dreams (BYD), which transitioned from manufacturing rechargeable batteries to becoming a major player in the electric- and gas-powered automotive industry.

Similarly, the Dafen Oil Paint Village, which continues to reproduce masterpieces by artists like Van Gogh, has emerged as a recognized rural cultural industry. Artists in Dafen work in small studios and contribute to a *Gesamtwerk*—a collective art production that preserves the skill of imitation as a cultural asset. The village's own museum honors this unique cultural design and celebrates the expertise involved in replicating the works of masters, thereby validating *Shānzhài* as a form of cultural and economic resilience (Landsberger 2019).

Another illustration includes the seven characteristics of traditional learning in calligraphy, in which DeCocker (1998) includes (1) copying the model and (2) nonverbal communication. He states that mastery of the model is of foremost importance, creative interpretations of which are discouraged. Imitation of the model is also encouraged because verbal communication is limited among master and apprentice.

The *Shānzhàides* of the world challenge our conventional notions of ownership, intellectual property, and the exclusive rights that accompany them. The idea of what constitutes "mine" is not universal but is instead shaped by cultural norms, legal frameworks, and economic systems that have evolved over time. For example, in the Edo period of Japan, the *Kanō* school of painting employed a method of instruction that revolved around the art of copying. Apprentices were trained to replicate faithful and rough, scaled-down reproductions of original works; these replications were known as *funpon* (copybooks). These *funpon* were created by the masters not just to teach their visual language but also to maintain and perpetuate the artistic style of the school. The secrets of the master's art were not disclosed through explicit instruction or by directly observing the master at work. Instead, they were embedded in the *funpon*, which served as both a pedagogical tool and a means of preserving the master's techniques for future generations. The practice of copying in the *Kanō* school was not merely an exercise in reproduction; it was a fundamental method for transmitting knowledge and ensuring the continuity of artistic traditions. Talent alone was not sufficient; what mattered was the ability to learn and reproduce through systematic training. This approach allowed the school to sustain elite patronage and manage its large enterprise, which ensured that its style and techniques were passed down through the generations (Kazuko 2014). Even the masters themselves engaged in copying the *funpons* of other masters to refine their skills further, which emphasizes the importance of this method for continuous learning and improvement.

So highly is the art of copying regarded that, since ancient times, bamboo and plum paintings made with ink are not identified as "bamboo painting" or "plum painting" but as "bamboo copy" or "plum copy." The artists copy the essence of bamboo and plum as their minds have captured it while observing the bamboo and plum (Kazuko 2014). Unlike the landscape paintings of the Romantic or Impressionist period that hailed autonomy, the Japanese art theory credited nature for permitting the artists to copy it.

More importantly, from a modern perspective of demand and supply, opening the know-how to all interested individuals equips the entire community with the skill set required to meet large production orders in the absence of tools of mass manufacturing. In the case of the *Banārasī* toy makers, the open emulation of skills and techniques within the community ensures that large production orders can be met even in the absence of mass manufacturing tools. A single toy maker would be unable to fulfill such orders alone, but through emulation, all members of the community can contribute to the production process, thereby meeting client demands and sustaining their livelihoods.

It is a misnomer to call some of the Indian cultural design communities small scale. Some of these communities drive the entire city's economy (for instance, brassware in Morādābād and *chikan* in Lucknow). Wilkinson-Weber and DeNicola (2020) highlight the *chikankārī* embroidery industry in Lucknow as a compelling example of how emulation operates as a legitimate and necessary production method in cultural design. In this context, *chikankārī* embroidery, predominantly carried out by women within the domestic space, connects to a broader network that extends beyond the home. This network includes men who facilitate various stages of production, such as transporting materials, printing, washing, and selling the finished products in showrooms across the city and beyond. Despite being labelled as a cultural design due to its handmade nature and lack of machine power, the *chikankārī* industry has an economic impact and scale of operations comparable to that of industrialized units.

The success and scale of the *chikankārī* industry are made possible through emulation, wherein embroiderers replicate the designs of their peers and clients within the confines of their homes. Such collective efforts enable communities to meet the substantial market demand for their products without relying on mass manufacturing technologies. Emulation, in this case, is not merely a means of production but a vital practice that allows for the transfer of knowledge, the preservation of generational skills, and the ability to adapt to market needs.

The re-positioning of cultural design in the modern context hinges on the emergence of new signifiers that reflect current societal values, technological advancements, and global interactions. These signifiers enable cultural design to transcend its traditional boundaries and engage with contemporary issues such as sustainability, identity, and social justice. By adapting to the modern context while maintaining its connection to tradition, cultural design can continue to thrive and offer a reflection on generational knowledge, a justification for its continuing practice, and a vision for the future.

References

Adamson G (2007) Thinking through craft. Berg, Oxford

Alexander C (1964) Notes on the synthesis of form. Harvard University Press

Bloom H (1997) The anxiety of influence: A theory of poetry. Oxford University Press, USA

Carbonell JG (1982) Experiential learning in anagogical problem solving. In: Proceedings of the second meeting of the American Association for Artificial Intelligence, Pittsburgh, PA, pp 168–171

Chauhan VK (2022) Banarasi Saree a part of rich Indian culture—a review. J Nat Fibers 19(17):15723–15737

Coomaraswamy AK (1909) The Indian craftsman, vol 1. Probsthain & Company

Crowder K (2016) Producing Suffolk punch horses: craftsmanship and sentient media. In: Craftwork as problem solving—ethnographic studies of design and making. Routledge, pp 51–71

DeCocker G (1998) Seven characteristics of a traditional Japanese approach for learning. In: Singleton J (ed) Learning in likely places: varieties of apprenticeship in Japan. Cambridge University Press, Cambridge, pp 68–85

Dreyfus H, Dreyfus SE (1986) Mind over machine. Simon and Schuster

Doboli A, Umbarkar A (2014) The role of precedents in increasing creativity during iterative design of electronic embedded systems. Des Stud 35(3):298–326

Eilouti BH (2009) Design knowledge recycling using precedent-based analysis and synthesis models. Des Stud 30(4):340–368

Ericsson KA, Krampe RT, Tesch-Römer C (1993) The role of deliberate practice in the acquisition of expert performance. Psychol Rev 100(3):363

Ertl R, Hibberd R (2003) The art of the Russian Matryoshka. Vernissage Press, Colorado

Gates D (2016) From 'in our houses' to 'the tool at hand': breaching normal procedural conditions in studio furniture making. In: Craftwork as problem solving—ethnographic studies of design and making. Routledge, pp 115–133

Greenhalgh P (2002) Introduction: craft in a changing world. The Persistence of Craft, London: A&C Black, pp 1–17

Guth CME (2010) The multiple modalities of the copy in traditional Japanese crafts. J Mod Craft 3(1):7–18

Hua Y (2011) China in ten words, 1st edn. Pantheon Books, New York

Kameda-Madar K (2014) Copying and theory in Edo-period Japan (1615–1868). Art history 37(4):708–727. https://doi.org/10.1111/1467-8365.12111

Landsberger S (2019) 2.5 Shanzhai = creativity, creativity = Shanzhai. In: de Kloet J, Chow YF, Scheen L (eds) Boredom, Shanzhai, and digitisation in the time of creative China. Amsterdam University Press, Amsterdam, pp 217–224. https://doi.org/10.1515/9789048535538-015

Latour B (2008) A cautious prometheus? A few steps toward a philosophy of design (with special attention to Peter Sloterdijk). In: Networks of design, annual international conference of the Design History Society. University College Falmouth, Cornwall, pp 2–10

Madeley C (2007) Copying cars: forgotten licensing agreements. In: The culture of copying in Japan. Routledge, pp 251–268

Marchand THJ (ed) (2016) Craftwork as problem solving: ethnographic studies of design and making. Ashgate Publishing, Ltd.

Nonaka I, Takeuchi H (1995) *The Knowledge Creating Company.* New York: Oxford University Press

Patil K, Uday A (2017) Mutable mobiles: charting the cartography of Banarasi toymakers. Design Issues 33(2):43–57

Patil K, Uday A (2019) Gestural drawings: a study of spatial data held in narrative design ges- tures. Int J Archit Spat Environ Des 13(1):35

Pye D (1968) The nature and art of workmanship. Cambridge University Press, Cambridge

Reitan JB (2006) Inuit vernacular design as a community of practice for learning. CoDesign 2(2):71–80

Rittel HWJ, Webber MM (1973) Dilemmas in a general theory of planning. Policy Sci 4(2):155–169

Schön D (1983) The reflective practitioner, how professionals think in action. Basic Books

Sennett R (2008) The craftsman. Yale University Press, New Haven

Sturt G (1963) The wheelwright's shop. Cambridge University Press

Wilkinson-Weber CM, DeNicola AO (eds) (2020) Critical craft: technology, globalization, and capitalism. Routledge

Yanagi M, Leach B (1989) The unknown craftsman: a Japanese insight into beauty. Kodansha International

Chapter 9
Designing "Others": Significance of the Hand to Practices of the Mind

1 The Zeitgeist of Cultural Design

As we conclude our study, which primarily focuses on cultural design thinking among the *Banārasī* wooden toy designers, *we* observe that parts of this study are also relevant to mainstream discussions of design thinking and cognition. In this concluding chapter, *we* situate cultural design thinking in the contemporary discourse and, in doing so, *we* attempt to relocate cultural design as an alternative practice and not as the archiving of a museumized past.

Many researchers believe that the tide has changed for the cultural design discourse. Cultural designs, designers, and design skills are experiencing a zeitgeist moment of popularity as they mark a turning point towards the tactile (Brink 2024; Groth et al. 2013). It is their "moment in the sun" (Luckman and Thomas 2018, 1). Though cultural design has often been considered primitive, pre-industrial, and unsuitable for a modern way of living, the emphasis on a maker's ability to think with their hands has reconfigured entire sectors in recent years. In *The Craftsman*, Sennett (2008) describes how cultural design is an enduring, basic human impulse—the desire to do a job well for its own sake. It is about the intrinsic satisfaction of doing good work, caring about detail, and making desirable products.

While the maker movement has significantly transformed our conception of and attitude toward DIY (do it yourself)—introducing a novel approach to technology through RPTs, small CNC machines, and other tools for smart fabrication by individuals rather than factories—cultural design workshops continue to be celebrated as its precursor (Hatch 2014). Unlike Mark Hatch's TechShop, where tools and manufacturing capabilities are distributed but not owned by the maker, cultural designers possess their means of production and have cultivated a robust community of practice and learning. This practice of decentralized ownership is largely unknown to mainstream maker communities.

Moreover, these erstwhile maker spaces represent a promising force for change, one that may facilitate the transition toward a sustainable society (Bell et al. 2018;

K. Patil, *Toying with the Mind*, Design Science and Innovation,
https://doi.org/10.1007/978-981-96-9227-9_9

Moore 2005). Cultural design practices utilize materials and appropriate technology, with consumption often limited to a specific context; thus, cultural design is frequently associated with a circular economy. With its small carbon footprint, cultural design serves as an alternative social innovation strategy (Bofylatos 2017; Sforzi and Colombo 2020)—a means of uplifting those at the Bottom of the Pyramid (Al-Dajani and Marlow 2013). Unlike the industrial owner of capital, cultural designers possess the will, values, goals, and means to develop and preserve the land and resources they occupy. They are both creators and consumers of the development that surrounds them. In this sense, cultural designers embody the ideal citizen.

Several scholars have also repositioned cultural design as a community of practice and learning (Gibson 2019; Lave and Wenger 1991) instead of a cottage industry employing slow means of production. The pedagogical discourse surrounding cultural designs now enriches the mainstream theory-based curriculum, with Scandinavian countries, particularly Finland, leading the way. Cultural design has been integrated into the Finnish school curriculum since 1866, where it functions as a learning-by-doing pedagogy rather than merely an extracurricular activity (Jaatinen and Lindfors 2019). Reitan (2014) strongly advocates for incorporating learning techniques derived from cultural design, such as observation, into design classrooms.

In addition to its role as a curricular aid, cultural design is increasingly prominent in the discourse on well-being. It is being examined as a therapeutic tool (Huotilainen et al. 2018) and as a means of subliminally engaging with materials and transforming emotional stress into feelings of relaxation and a sense of empowerment (Collier 2011; Maidment and Macfarlane 2013; Pöllänen 2015; Reynolds 2000).

The rebirth of cultural design is also strongly acknowledged and affirmed within academic circles. Malcolm Ferris, in his Making Futures conference, which he curated from 2009 to 2021, positioned cultural design as a vision for the future. Convinced of the transformative potential of small-scale making and its capacity to contribute to new progressive futures, the conference submissions explored contemporary cultural design and maker movements as "change agents" within society in the twenty-first century, particularly concerning global sustainability agendas, social entrepreneurialism, and community regeneration. The *Craft Research Journal* seeks to portray and establish cultural design as a vital and viable modern discipline with a forward-looking vision. Similarly, the *Journal of Modern Craft* conceives of cultural design as both an idea and a field of practice in its own right and reviews and analyses its relevance within new media, folk art, architecture, design, contemporary art, and other related fields.

Cultural designs are experiencing a resurgence and are being rediscovered as both a hobby and a sought-after enterprise (Jakob 2013, 127; McIntyre 2010). This practice has proven to be a viable business venture, emerging as a thriving multibillion-dollar industry during the global economic recession (Jakob 2013, 127). According to the Indian Exports Promotion Council of Handicrafts (EPCH), the exports of handicrafts were merely ₹386.57 crores in 1986–1987, but this figure rose dramatically to ₹25,679.98 crores in 2020–2021 ("About EPCH—Promoting and Facilitating the Export of Indian Handicrafts Across the Globe" n.d.).

In this chapter, we review these contemporary notions that break the myth of cultural design as fossilized remains that are redundant in the modern scenario. Of course, our discourse pivots around the *Banārasī Khilonas*.

2 Cultural Design Methods of Problem Solving

Tipu's toy is an engaging eighteenth century musical hand-crafted automaton designed by unknown wooden toy makers from India, probably in collaboration with French watchmakers. The tiger's carved and painted wooden body is seen mauling a British soldier and is currently a popular attraction in the Victoria and Albert Museum in London. According to Henry Willis,

> The only moving part is the left arm of the victim which oscillates between the victim's mouth and the tiger's ear. Otherwise, apart from the figures themselves, this ingenious apparatus obtains its realism from the sounds emitted. These sounds are of great interest to organ-builders since their source is a semi-mechanical pipe organ within the tiger and the victim, operated by a cranked handle from the left of the tiger's chest. The handle operates a worm gear and crank-shaft to the three sections of the organ. (as cited in Mildred 1959)

The narrative surrounding the origin and creation of certain historical automata, particularly the one described, illustrates the complexity and collaborative nature of cultural design. While some argue that the mechanisms and parts of this toy may have Western origins—citing records of French watchmakers, designers, and engineers in Seringapatam (Mildred 1959; Stronge 2009)—the true origins of the design remain debated. The possibility that a French organ maker might have contributed to the design cannot be ruled out, yet no definitive evidence establishes a clear role for the French in the creation of these automata.

What is evident, however, is the significant influence of Indian cultural designers, particularly the *Chitrakārs*, in the creation of these intricate toys. The visual representation techniques and motifs employed in the automaton are consistent with those used in other toys from the same period. The use of round, lathe-turned tops in the keyboard instead of conventional keys is a clear indicator of the *Chitrakārs*' artisanship, which can still be observed in the lathe-turned toys made in Channapatna today.

The design and construction of this automaton, with its intricate mechanism of musical performance integrated into the body of a tiger and synchronized with its movements, presented an extraordinary challenge. The complexity was further heightened by the need for collaboration between Indian designers and possibly French experts, despite the barriers of language and differing design philosophies.

The automaton stands as a remarkable example of systemic and collaborative design and demonstrates the ingenuity and problem-solving capabilities inherent in the cultural design process. Mastery, here, is the ability of toy designers to navigate between and outside the categories of familiar and unfamiliar objects and skills and creatively translate knowledge from one context to solve problems in another—from watchmaking to toy automatas.

The examples of problem solving drawn from the *Banārasī* community of toy designers (Patil and Athavankar 2022), the instance of Tipu’s Tiger discussed above, case studies of the Suffolk punch horses, bike mechanic’s workshop, furniture making, weaving, and ceramics narrated from the perspective of craftwork as problem solving and discussed in the book edited by Trevor Marchand (2016); the longitudinal study of Vai and Gola tailors (Lave and Wenger 1991); the anthropological accounts of the Hopi basket weavers (Teiwes 1996); accounts of technological and social innovations in English farmhouse cheese making (Blundel and Smith 2013); and several others in the literature suggest that cultural designers do solve problems. Marchand (2016) argues that calculation, theorizing, setting goals, imagining outcomes, and working out hypothetical pathways toward a solution are very much a part of both modern design and making in cultural design. As such, cultural designers solve problems, but in a way that differs from how modern designers solve problems.

In exploring the nature of problem solving, the cultural design discourse has reached a consensus on several key aspects. Problem solving in cultural design is deeply embedded within its context—situated within the specific environment, social structures, and professional communities that influence the designer’s process. Cultural designers engage with a network of other professionals, social actors, and the environment, all of whom pursue gains in knowledge, social identity, and professional status through their work.

Cultural design problems frequently arise in tandem with the process of self-evaluation, where designers identify mistakes and opportunities for improvement while mastering a tool or technique, improvising their skills, or innovating to meet the demands of a client. These problems are often the result of boundary conditions imposed by rigid traditions, which paradoxically fuel a constant need for incremental changes. The immediacy of the solutions, rooted within a cultural framework, often results in problems that are closely tied to local resources, practices, and technology.

The design process within this context is evolutionary rather than revolutionary. It involves small, incremental changes that ensure a good fit between the evolving form and its changing context. This approach acknowledges that mistakes and failures are integral to the learning and innovation process and lead to incremental improvements in both the designers’ know-how and the artefacts they produce.

Alexander (1964), among others, has articulated this idea and suggests that the problem-solving process in cultural design is characterized by its ability to maintain an evolutionary equilibrium—a balance between form and context that is continually adjusted through small, incremental changes. This method of problem solving is inherently conservative and aims to preserve the good fit achieved through tradition while allowing for gradual adaptation. Hence, problem solving is considered an ongoing, continuous activity and not a response to an exceptional event (Alexander 1964; Jones 1970; Marchand 2016; Bunn 2022). The resulting work is an incremental change in the know-how of the practitioner and the preceding artefact on which the innovation is built. This makes cultural design problem solving an iterative process with a failure-positive approach (Martin 2015; Sheridan et al. 2014).

However, this incremental nature of problem solving in cultural design raises important questions about the scope, scale, and complexity of the problems being addressed. Unlike modern designers, who may tackle large-scale, complex challenges requiring radical innovation, cultural designers often deal with problems that are more localized and immediate. It can be deduced that cultural designers do solve problems, albeit the nature of these problems does not have the same scope, scale, and complexity as that of the modern designer. What exactly are cultural design problems? The following sections discuss characteristics of such problems.

2.1 Problems of the Body and the Senses

Material and how it speaks back to the designer has often been a legitimate inquiry in the design literature till late. However, increasingly, the design process is skipping the traditional steps of learning and unlearning as computers mediate between the designer's thinking and the material. Therefore, SOLIDWORKS (a software used to create 3D CAD models and assemblies utilizing a parametric feature-based approach) responds to the pinch or finish that the designer decides to apply to the material, which eliminates the need for touch and real-time feedback on the material. Artificial intelligence (AI) is also being employed to skip steps in the design process related to concept generation, data collection and analysis, and visual representations of the same. On the contrary, the cultural design process remains entrenched in the experience of the materials and its processes.

In cultural design, the relationship between the body, its sensory organs, and the material is intimate and unique and forms the foundation for the mastery of cultural design. A master weaver, for instance, can discern the type and extent of an error on the loom merely by the sound it produces. Similarly, a dye maker can taste and judge the final color of a dye, while a potter senses even the slightest change in clay as influenced by the wheel's speed. This heightened bodily awareness and sensory acuity are crucial in cultural design, where mastery is not only a matter of technical skill but also of an embodied understanding of the material, tools, and environment.

The concept of mastery in cultural design can be understood, in part, by assessing how a designer overcomes the limitations of the human body through a continuous dialogue with materials, tools, and surroundings. The challenge lies in pushing the body beyond its natural threshold—whether that means enduring longer hours at the loom, seeing and placing minute beads in a *zardosi* brocade, holding the hand stable while drawing each strand of hair in a visual, sensitizing the hands to know the behaviour of clay, reading the mood of a horse by sensing its skin, or speeding up the process of beating the dough in the *mochitsuki* process. These are not merely tasks; they are bodily habits cultivated through persistent practice and emulation.

Problem solving in cultural design, then, is inherently a problem of shaping and maneuvering the body to effectively interact with materials and processes. This involves two types of bodily habits. The first kind involves the familiar or typical artefacts which the body has practiced and trained itself upon. Frequent and constant emulation makes these bodily habits tacit. However, like Nakatanidou, who strove to outperform himself in the act of pounding the glutinous rice, cultural designers are constantly emulating themselves to overcome the limitations of their bodies. They challenge their bodies to perform actions that are more precise, more sensitive, more delicate, faster, more slow, more strong, more rhythmic, or more synchronized with material, tools, and processes. The designer's body becomes a critical tool in itself, one that must be trained, disciplined, and attuned to the demands of cultural design.

The second type of problem solving in cultural design involves the body learning new ways of moving, sensing, and acting to create an unfamiliar artefact. Unlike the first kind, where the body relies on ingrained habits and tacit knowledge, this second type demands the development of new bodily habits and skills. These non-habitual actions necessitate a fresh learning cycle, which can be both facilitated and hindered by existing bodily habits.

When confronted with a new design challenge, the cultural designer's body must adapt to unfamiliar movements, sensations, and techniques. This process is often complex because the established habits of the body, formed through years of practice, may interfere with the learning of new actions. The body may resist the new motions and revert to familiar patterns, which can slow down or complicate the learning process. However, these older habits can also provide a foundation on which new skills are built because cultural designers draw on past experiences to inform their approach to the new task.

The challenge in this second kind of problem solving lies in the need for sustained effort to reach the body's current limits and then push beyond them to establish new capabilities. The process is iterative and gradual and requires cultural designers to repeatedly challenge their bodies to expand their range of motion, sensitivity, and control. This continuous cycle of pushing against the body's limits, adapting, and refining actions is essential for mastering the creation of unfamiliar artifacts (refer to Fig. 1).

In body-dependent problem solving, tools function as natural extensions of the human body, enhancing its ability to meet the challenges posed by material and design processes. These tools do not displace or replace the body; rather, they are integral in helping the body overcome its limitations. The tools amplify the body's

Fig. 1 The cycle of body limits. *Image source* Author

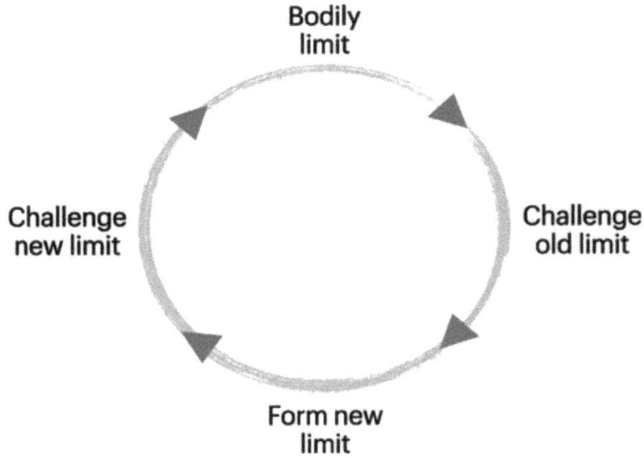

strengths and allow cultural designers to push beyond their physical boundaries and achieve feats that would be impossible using the body alone.

2.2 *Problems of Approximation and Judgment*

Scholars of modern design agree that contemporary designers aim to achieve the adequate rather than the comprehensive (Nelson and Stolterman 2014, 99), the satisficing rather than the optimized (Simon 1969), or the culturally appropriate rather than universally acceptable (Friedman 2005). And despite being moderated by the mean, the discipline is often veered toward outliers in the unending quest for novelty and newness. *Adequate*, *approximate*, and *satisfying* are often used as synonyms in modern design to indicate the inability of the design profession to meet all the requirements of all users. Instead, the profession aims to satisfy some of the requirements of some of the users. Several tools, such as TRIZ (Theory of Inventive Problem Solving[1]) and QFD (Quality Function Deployment[2]), are employed to evaluate design trade-offs and select the parameters most essential to the design solution and to the user. The parameters or functions lower in the hierarchy are removed from the design considerations. Similarly, the mean set of users and their characteristics alone are designed for. Within this subset, some characteristics are accepted to be only partially fulfilled—for instance, the comfort of a chair cannot be achieved completely. In modern design, satisficing, then, is a cognitive heuristic that necessitates skimming through the available options or alternatives until an acceptability threshold is met (Colman 2006, 670).

> One of the modern design's key distinctions is that design choices are made through design judgments, leading to the creation of something that did not exist before. As such there is little gain in directly copying or imitating earlier designs. There is no need to survey other designs, with any other purpose than to influence or stimulate creativity and catch a sense of the method of designing as an activity, unless it serves the purpose of historical or critical interests. (Nelson and Stolterman 2014, 99)

On the contrary, in the cultural design discipline, an approximation is the acumen gathered from the palimpsest of experience obtained by rehearsing particular actions, in a specific sequence, over a selected material through the engagement of definite tools to achieve certain consequences. This process of knowledge generation is not irrational or unreasonable but intuitive and tacit. It is not employed to design adequate features for the mean population, but to strive repeatedly to imitate an idealized form that has been set as the goal by the collective design mind. Approximate decision

[1] Theory of Inventive Problem Solving or TRIZ was developed by a Russian engineer—Genrich Altshuller and his colleagues from the study of over 400,000 technology patents. They drew basic patterns of problem solving to create new ideas and innovation from these patents.

[2] Quality Function Deployment or QFD is a systematic management tool to transform customer needs into design functions that ensure quality parameters through the designing of components and sub-systems.

making and approximate construction of the material do not imply a lack of precision in manifesting the structure, dimensions, and function of the paper sketch into a physical prototype. In contrast, it is a powerful strategy of continuous learning and iteration to refine and enhance the designer's attempt to resemble the best fit to the context.

For cycles of approximations to be undertaken, the cultural designer must have the freedom to approximate and an environment that honors and encourages approximations. The expectation that perfection will be achieved in any attempt, even after multiple tries, can kill the cycles of approximation. In the traditional methods of designing, it is well accepted that approximations are the will to do a job better than the previous attempt, not with the ego that perfection will be achieved with the next attempt, but with the humility that this elusive goal will always be beyond reach. Arguably, no one will attempt to approximate if the risk of the attempt or its output is demeaned by the community (Cambourne 2022, 61). In cultural design communities, approximating is an honorable and legitimate method of learning, iterating, and improvising the skill and its material consequence. Therefore, the problem statement remains the same, but the problem of approximating the solution is shared over generations.

The generational nature of the cycles of approximation increases the scope of the problem of approximation. As apprentices mature and eventually become masters themselves, they inherit the responsibility of transmitting these approximation cycles to their own apprentices, as depicted in Fig. 2. This process establishes a generational chain (Generation 1, 2, *n*, etc.), where each generation contributes to the refinement of the design. Over time, the number of approximations typically decreases as the design stabilizes and becomes a standard or typical artefact within the cultural system. The need for further approximations diminishes, either ceasing or becoming infrequent as the design evolves into a widely accepted and replicated typical instance within the community.

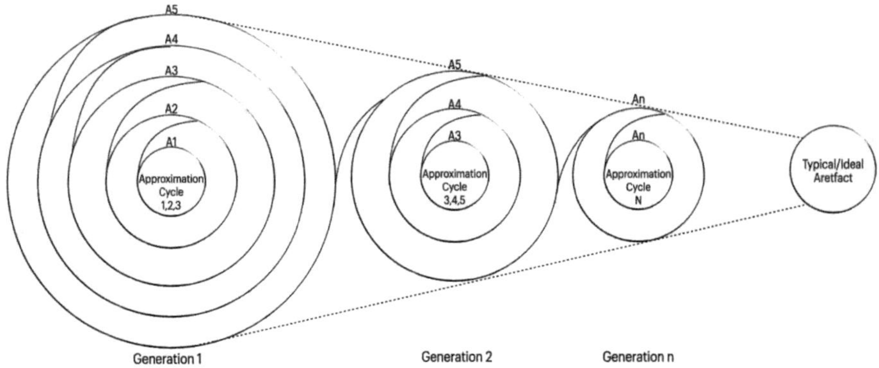

Fig. 2 Generational chain of approximation cycles that leads to the best typical fit to the context. *Image source* Author

However, this continuity and refinement of approximations presents a challenge, particularly in contexts where the knowledge transfer is predominantly oral and unrecorded in writing or sketches. In many traditional Indian contexts, for instance, the process of design refinement relies heavily on indigenous tools that facilitate the handling and transmission of information within the community. These tools include oral language systems that support shared categories, customs, and conventions of making artefacts; nonverbal methods such as gestures, observation, emulation, and learning by doing; and, in literate communities, documentation through texts and diagrams. These practices support the sustained evolution of cultural designs and ensure that each generation can build on the approximations of its predecessors, refining and perfecting the design until it becomes a stable and recognized part of the cultural system.

Approximation in cultural design is often driven by self-will and intrinsic motivation rather than external pressure or structured guidance. This internal drive is crucial because a collective, synchronized effort from the entire community of stakeholders is required to move toward achieving a better-fit solution, if not the ultimate best fit. However, this process of collective effort and communication becomes more challenging in cultural design communities that are informally organized and educated, as is the case in India and elsewhere.

Cultural design communities informally learn the trade and organize the manufacturing and assembling of their products without explicit coordination and transfer of decision making. This makes the process of approximation fragmented and scattered between several *kārkhānās*. Without explicit communication and coordination, it is challenging to ensure that each individual approximation aligns with the broader goals of the community or the overall design vision. The fragmented nature of the process can result in inconsistencies, with different parts of the product or design evolving in divergent directions. Making a meaningful whole out of such disparate improvizations and innovations is a continuous struggle in the cultural design process.

2.3 Problems of Thinking-Enabled Production

Cultural designers engage deeply with tools and materials, where sensing, doing, and making are integral to their thinking and problem-solving processes. In cultural design, the problem itself is inherently defined by the material and the tools at hand. For instance, *Banārasī* toy designers, who work primarily with wood, frame all their design problems around the material properties and the tools they use, such as turning, cutting, and carving. These designers do not engage with problems involving plastics because their expertise and material consciousness are rooted in woodcraft. Similarly, a weaver's design challenges are confined to textiles shaped by the loom, while an equine trainer's problem solving revolves around taming and disciplining horses, not other animals like elephants.

Primarily, cultural design challenges remain a form and context problem that must be solved with a particular material and skill set. The material, in cultural design processes, literally speaks back and comes alive in the process of problem solving. The inherent characteristics and limitations of the material and the tools shape the solution. Sennett (2008, 119–148) describes this "material consciousness" as an attribute of any work built on the curiosity of the material in hand and extends cultural design skills to the precise and more or less explicit practice of surgery at the hands of a surgeon. He looks at surgeons as cultural designers and the audience of the medical community watching them as members who share the same material consciousness. The surgeon thinks through her hands, which display a practiced skill and the ability to improvise her actions in unpredictable circumstances. In this sense, we can view surgery as a cultural design practice that evolves on the operating table—analogous to how changes in cultural design often occur directly on the factory floor or in the *kārkhānā* rather than on a pre-planned sketch. The mind of the surgeon and the cultural designer is expressed in the hand; therefore, cultural design methods of problem solving, like any other process of the mind, are of significance to contemporary studies of the mind and its process of thinking in the making and changing of any artificial situation or artefact.

Unlike modern design practices, which often rely on detailed planning, user surveys, market benchmarking, and other sources of insight to inform design decisions, cultural design thinking happens in real time as the hand engages with the material. This means that design and production occur simultaneously and form a reflective practice where the act of making is an ongoing process of learning and adaptation. Such thinking-enabled production assists in worldview acquisition through the knowledge of available material, technology, and skill. Here, the designer works within the known—what is already understood about the material and tools—to explore the unknown and arrives at solutions through a process of combining and experimenting with what is at hand.

On the other hand, modern design practices, which often start with designing on paper, involve a more speculative approach. Designers begin with an informed conjecture about what the design could be based on data from users, markets, and production capabilities. This conjecture represents the unknown—a hypothesis about how the design might fulfill a need or solve a problem—while the designer works toward the known—refining the design within the constraints and possibilities that emerge.

Dorst (2011) discusses a similar concept, where designers, in their pursuit of an aspired value, propose a framework or working principle that guides the design process. This proposed framework is speculative and cannot be deemed definitive until the design is fully realized and proven to achieve the desired outcome. In this sense, modern design thinking often moves from the unknown (a conjecture) toward the known (a validated solution), whereas cultural design thinking moves from the known (material knowledge and skills) toward the unknown (a solution that emerges through making). While both domains involve creativity and artisanship, the focus and approach in cultural design are distinct, particularly in how problems are defined and addressed.

In cultural design, problems are deeply rooted in the material and the tools available to the designer; therefore, cultural design does not always generate disruptive innovation. In the medical field, for example, surgeons might use the same set of tools and methods throughout their career, with technological changes occurring gradually and often in alignment with existing practices. This reflects the iterative, evolutionary nature of cultural design, where innovations are incremental and built on established methods and materials. On the other hand, modern design problems are often not shaped by the material or the tool but by the characteristics of the product. Therefore, the modern designer will not be tasked with a problem in a specific material—say, cloth—but with the attribute of the product—say, flexible, eco-friendly, or plush—which may suggest cloth, vinyl, Minky, or synthetic fibres like polyester as a possible material to the designer.

In contrast, the cultural design process is rooted in the material and tools. For instance, if the material at hand is marble, the challenge lies in manipulating marble to achieve the desired attributes in the final product, such as making marble appear translucent or light. The process is highly material centric, with the designer's skill in working with that material being paramount. The work of Giuseppe Sanmartino, particularly his *Veiled Christ* (1753), exemplifies the challenge of imitating cloth in marble. Sanmartino's mastery of marble allowed him to sculpt the material to mimic the delicate, flowing qualities of cloth, thereby showcasing the level of skill and practice required to achieve such an effect.

Although the problems of material and tools are pervasive in cultural design, on rare occasions, the cultural designer is also expected to employ a well-rehearsed sensibility of one material on another. Doing so involves cycles of making that continuously refine and improve the designer's skill and knowledge. The coupling of learning with making on the factory floor is a key aspect of cultural design. A notable example of this adaptability can be seen in the work done by female Navajo seamstresses for NASA during the *Apollo 11* space mission in 1969; this work continued until integrated chip design replaced the core rope memory. At its peak, about 1000 Navajo women skilled in weaving rugs used their knowledge of thread and rope manipulation to interconnect tiny chips with wires under a microscope. The seamstresses, with nimble fingers and an infallible eye for detail, threaded metal wires through tiny doughnut-shaped ferrite rings, or "cores," that represented binary numbers. With the humble wire-threaded needle, the core memory of one of the most sophisticated engineering problems was woven. The abstract patterns the Navajo women created in this high-tech environment mirrored the designs they traditionally wove and showcased their ability to apply a well-rehearsed sensibility from one material (threads in weaving) to another (wires in microchips). This required the women to transfer their weaving knowledge of ropes and threads to that of electronic wires (refer to Fig. 3). Similarly, when engineers could not coat Teflon, Lien Pham, a Vietnamese seamstress seeking refuge in the United States, suggested folding Teflon at the edges of the object and sewing it like a hem. The analogy worked. Women like Lien Pham, employed by NASA in its space program, were called *LOL*—Little Old Ladies—who helped, literally, sew the entire code of the Apollo Guidance Computer.

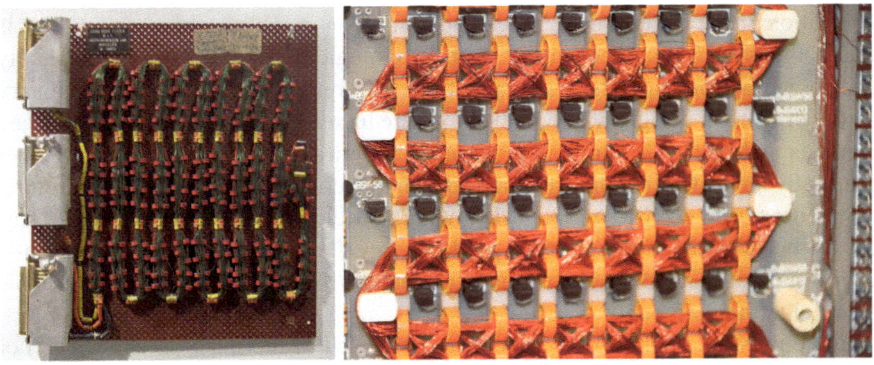

(a) (b)

Fig. 3 **a** Ferrite core memory as used in the Apollo Guidance Computer. **b** A 16-Kb ROM board: Wagner 02 MIS 01 5001 X from a Wagner WAC-40 or WAC-400 computer, 1974. *Image source* Wikipedia—These two files are made available under the Creative Commons CC0 1.0 Universal Public Domain Dedication. https://commons.wikimedia.org/wiki/File:16Kb_Rope_Core_Memory_Board_from_Wagner_WAC-40.jpg. https://en.wikipedia.org/wiki/Core_rope_memory#/media/File:Apollo_guidiance_computer_ferrit_core_memory.jpg

2.4 Problems of Making Recurrent, Habitual Actions

Neuroscience traces our habit-making tendency to the basal ganglia in the brain, which plays a key role in the functioning of our voluntary movements. The nerves and neurons of the basal ganglia also process the signals received from our sensory organs—namely, sight, sound, smell, taste, and touch—which further refine the body's response to stimuli or problems. Frequent exposure of the basal ganglia to an activity or object forms a habit.

According to Duhigg (2012), every habit is a three-part process that eventually forms a psychological pattern called a "habit loop." The loop begins with a cue or a trigger that unfolds a behavior in response to the situation or object. If the exposure to the situation or object is frequent, a routine response is created. To repeat this response, the brain requires a "carrot," which is a positive affirmation of the action that helps it to store and retrieve this action every time, thereby forming a habit loop.

The habit loops of cultural design, which are characterized by their iterative, hands-on problem-solving approach, stand in stark contrast to the emphasis on novelty and disposability in modern design solutions. In modern design, however, the drive for newness and the pressure to produce marketable, innovative products often result in solutions that are not designed to be iterated upon after their initial launch. Once a product hits the market, the expectation is that it will meet consumer needs perfectly. If a defect or flaw is discovered, the solution is typically not revisited for refinement; instead, the entire consignment may be retracted or withdrawn from the market. This practice not only incurs significant financial losses for companies—affecting their balance sheets and potentially leading to costly recalls—but it also erodes consumer trust. In a marketplace driven by consumer loyalty, any perceived flaw in a product can create lasting suspicion. Consumers may become wary, question the reliability of the brand, and hesitate to exhibit loyalty in future purchases. Rittel and Webber (1973) draw the example of the design of a building that cannot be dismantled because it does not meet the required expectations. Buchanan concurs that solving a wicked problem is a one-shot operation that leaves no room for trial and error (1992, 16). This is unlike science, which often gives the scientist an opportunity to conduct several experiments to understand a phenomenon, and though each experiment has an economic cost to the scientific organization or lab and requires human effort, it has no social cost.

Similarly, the cultural design process is not a one-shot operation but a generational inquiry into bodily movements, materials, and tools. The outcome of such a process is not a singular, individual attempt but a palimpsest of actions that forms a generational habit inherited from parent to child. Every cycle of approximation is habit forming and feeds back into the evolutionary process. These habits are difficult to reverse because this knowledge and these skills are intergenerational and, therefore, cannot be wiped out. In addition, tacit knowledge is hard wired in the basal ganglia, which categories them as automatic actions. The brain is constantly searching for pathways of least resistance, and automating actions is an evolutionarily fit process for the same. Such automatic actions are difficult to decode from the brain, as seen in the classic example of HM.[3]

The neuroscience studies indicate that solutions are derived from habits, and new actions can be employed when they are beyond the scope of the existing habits. However, in cultural design, new actions have to become not only the habit of the individual designer but also that of the entire community, through which they become a part of generational habits. The design innovation in the flying toy by Ratanlal Prajapati was discussed earlier. In addition to the act of designing the flying toy is the process of how this design was disseminated in several *kārkhānās* and shops later. Merely a few weeks after the client received the beautifully carved flying toy, the new multiple-axis movement design was being made in several *kārkhānās* and sold in local shops. Ratanlal explained how this diffusion happened:

Author: I saw the new flying toy in Beharilal's toy shop.
Ratanlal: Ah, that one. He got it made from another carver.
Author: Really! The carver understood the new mechanism rather quickly, didn't he?
Ratanlal: The carver had come down here. I gave him the idea. He already knew how to carve the figures. He only had to carve them in different modules. Once done, I assembled that toy for him.

[3] After a surgical process in 1953, Dr. William Scoville removed parts of the medial temporal lobes, including the hippocampus, amygdala, and surrounding cortical areas, in an attempt to alleviate severe epilepsy suffered by patient HM. Unfortunately, as a result, HM developed severe anterograde amnesia; that is, he lost his declarative memory. But, surprisingly, HM could still master new motor skills such as mirror drawing, which indicates that procedural memory, or his ability to learn skills and create habits, was unaffected.

Author: And the painting? Painting was done by another toy maker; after all, everyone has their own skills.

If we outline the task flow of designing this new toy, we find that the design process involved multiple skill sets and collaboration with multiple designers in the absence of any plan or written instructions. Such complex negotiations or organizational skills are not often attributed to cultural design communities. However, Ratanlal suggests that such collaboration is an ordinary habit of action that is familiar to all toy designers in the community. Neither Ratanlal nor the other toy makers were willing to compete for client orders that were pouring in or claim the new design as their own. The design was in the open-source domain of the community. How are these design habits formed?

Similar to Gladwell[4] (2006) and Duhigg[5] (2012), habit loops in cultural design communities also begin with a person or a few people with particularly distinct characteristics. These initiators are members of a strong professional network of *kārkhānās* in which other skilled community members also participate. They have large families that sustain the master–apprentice culture and enjoy good market linkages with patrons. They are risk bearers who are willing to explore and experiment on behalf of the community at a faster and more disruptive pace than other members. The other members generally follow or support these risk bearers.

If the risks taken by these initiators are successful, then the innovation can be supported in two ways. One, the market may step in due to the support the risk bearers enjoy and place large orders, thus increasing the viability of the modifications. Large orders cannot be completed by a single or even a small group of cultural designers because cultural design communities do not have technologies for mass production. Therefore, the community pools its efforts to meet the market requirement and, in the process, practices making the innovation. Alternatively, the community members may appreciate the innovation and adopt it even before the market places an order. In this case, the entire community is prepared to meet the market demand.

However, market viability is not the only measure of success in making a habit in cultural design. For any change to be sustained in the cultural design system, the

[4] Gladwell (2006) compares ideas, artefacts, behavior, and messages to viruses. They spread like an epidemic in an optimal or suitable environment. These epidemics are controlled by three laws. The law of the few states that the success of any social epidemic depends on the involvement of people with a particular and rare set of social gifts. *Connectors* are the people who are social equivalents of a computer network hub. They know the largest number of people in the community across diverse social, cultural, professional, and economic circles. *Mavens* are reliable "information specialists" who connect us with new information. Salespeople are charismatic people who can persuade others with powerful negotiation skills. The second law discusses the stickiness factor of information or an object. It refers to the specific characteristic of the message that fortifies its influence on the receiver manifold, such that it "sticks" to the person. The last law focuses on the power of the context or the environment in shaping human behavior, such as punitive mechanisms.

[5] It is argued that social movements are triggered and maintained by three steps: (1) Movements are instigated by the social interlinkages of friendship, family, and the strong ties between close acquaintances. (2) Movements grow because of the habits of a community and the weak ties that hold neighbourhoods together. (3) Movements are sustained because leaders offer participants novel habits to own and identify with. This is the power of social habits (Duhigg 2012).

change must enter the learning cycle of the community. The innovation or modifications suggested by the risk bearers must become a part of the learner–apprentice loop. It is only when a change enters the cultural design pedagogy that a new feature, process, or function survives in a community. It is the master–apprentice transfer of information and skills that completes and sustains a habit loop over generations (refer to Fig. 4).

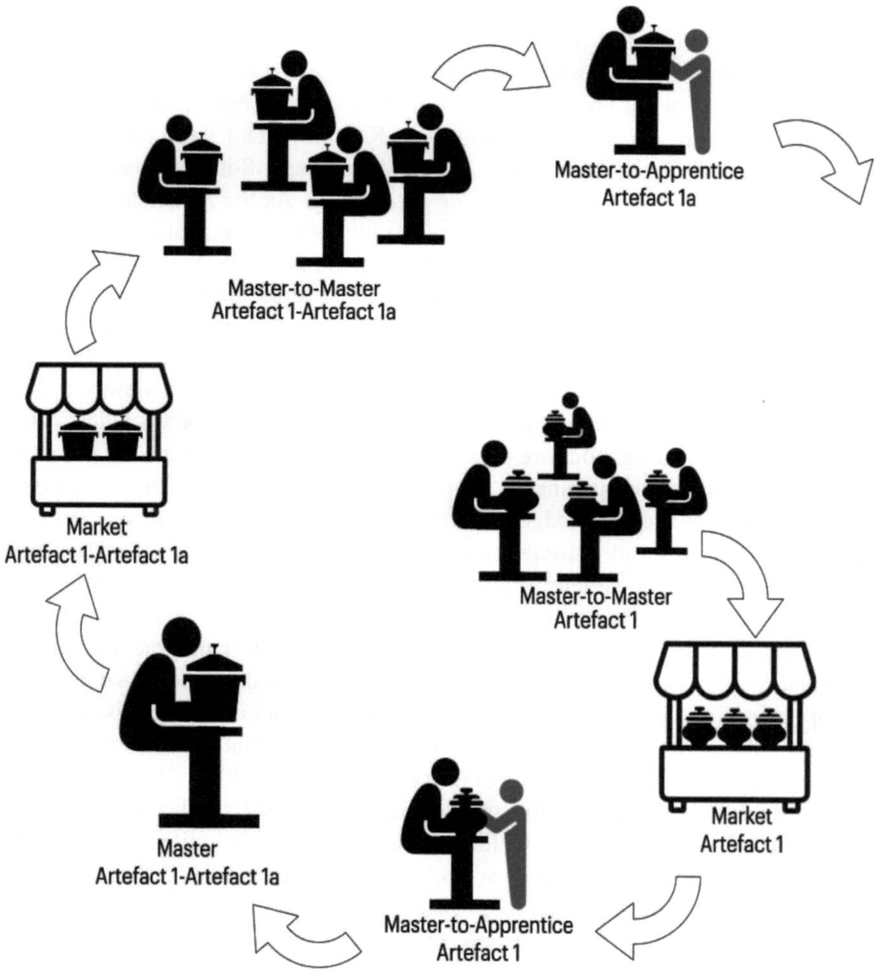

Fig. 4 Habit loop of cultural design communities. *Image source* Author

2.5 Problems of Assimilative Processes

Different kinds of cultural designs require appropriate training and methods of engagement, feedback, and problem solving (Gowlland 2016 in Marchand 2016). There is no fixed singular method or process like the double-diamond model, user-centric design, or steps to follow (e.g., empathise, define, ideate, prototype, and test) that will guide all cultural designers toward a more or less successful product. Every cultural design and its community of practitioners evolve local and contextual processes that fit the needs of the body, local material, and indigenous technology. Even though design skills like basketry, pottery, or weaving have maintained cultural continuities in every culture, they remain distinct in their representation, function, material, and form. *Jōmon* potters coil clay, Korean potters throw the clay on a wheel, and the Salmara[6] women beat the clay into shape. Some sun-bake the clay whereas others do biscuit firing, *raku*, or glaze firing. As the name suggests, blue pottery decorates the pots with blue glaze. The Jun kiln in China take pride in using red underglaze on pots, while the Nizamabad potters ingeniously use goat dung to achieve a lustrous black coating on their wares. Each culture finds its own continuity of distinct processes, which makes it difficult to make a generalized proposition. Moreover, these methods and processes are not hygienic (Law 2004, 9); they have biases, cannot be verified or replicated in altered conditions (worldviews), are not generalizable, and often do not have a reasonable explanation. Nevertheless, these diverse and divergent methods are generative in nature and, therefore, able to shape the production of an unimaginable set of artefacts. They reliably retain, transfer, assimilate, and change knowledge. It is important to divest cultural design from the singularity of definite and restricted sets of methods that claim superiority over other processes because they adhere to a set of determinate phenomena (user, ecology, gender, and purchasing power are examples). Instead of apologizing for the lack of rigor or the explicit nature of knowledge generation in alternate practices such as cultural design, Celia Lury and Nina Wakeford position these methods as enablers of the happenings of the social world. Within the social fabric, it is imperative to investigate how these methods legitimise the ongoing nature, relationality, contingency, and sensuousness of these happenings (Lury and Wakeford 2012, 2).

In cultural design, solutions to a problem are enumerable, unlike the wickedness associated with design problems. Cultural design problems are limited to the possible combinations of materials, bodies, and tools that fit into the context. These solutions within the boundary of combinations are acceptable to the community and explained by the *Weltanschauung* (worldview) of the designer. It is not possible to apply a method in cultural design that is indifferent to the context in which a problem is formulated; rather, it is specific and emerges through the problem at hand.

---

[6] Female potters of Salmara and their families inhabit the largest riverine island in Asia: Majuli. They do not use a potter's wheel. Instead, they beat the clay into shape, and finally polish the clay pieces. This technique can be traced back to the Bronze Age Harappan civilization. Archaeologists have hypothesized that Salmaran pottery could connect Mohenjodaro and Harappan civilisations to today's world.

These processes are appropriate representations of what Ferris (2009) and Silliman (2009) refer to as "changing continuities," or processes of continuity through change. The persistence of a native cultural design through negotiation, reinterpretation, and integration of cultural knowledge and external influences from other communities and markets has ensured the continuity of cultural design. Hence, they are assimilative methods of problem solving—an effort to approximate the archetype and not to innovate for the sake of novelty.

These assimilative processes bunch together individual patterns, categories, types, sequences of tasks, and combinations of actions, similar to how a database groups information; these groups form the generational matrix of knowledge and skills in a traditional community. Similar to artificial intelligence, which responds when prompted based on the data it is fed, cultural design is also as inventive as the boundary in which it is spread. In this sense, inventiveness is not intrinsic to the methods; rather, it emerges in relation to the purposes for which they are put (Lury and Wakeford 2012). As cultural design aims to maintain continuities of body, material, and tools, the method draws on several influences within and outside the category of objects to devise a solution. Assimilation of thoughts, processes, functions, forms, and materials into the habits of the body is imperative for the richness of the culture and its survival.

Change is not only a result of failure, as Alexander (1964) would suggest, for an un-self-conscious designer, but also a response to an external nudge with which a designer wishes to engage, either for economic reasons or for creative pleasure. However, the innovation of an assimilative practice does not end with creating a single artefact. Similar to habitual action, for any assimilation to survive, the process must be emulated often enough for it to become a habit of not only an individual but of the community, too. The challenge for assimilative practices lies in the identification of the appropriate sources of assimilation, the adaptability of the source to the process of assimilation, and the longitudinal span of integrating the processes within the community worldview. Assimilative processes can be said to be fully integrated into a community practice when the process of assimilation becomes a part of the learning trajectory of the apprentices or when the product of assimilation is christened in the local language or system of communication. Pecking toys renamed *Chuġtī Chiṛiyā* or *Hilāntā* and Russian dolls called *Sukhī Parivār* in the *Banārasī* dialect illustrate such assimilative practices.

3 Method of Self-organization

Zoybar is an open platform that leverages the collective intelligence of its user community to co-create virtual and physical musical instruments. By building on the technical expertise of its users, Zoybar encourages the improvization and sharing of production files within the platform. The shared modular kit, which consists of individual parts, can be modified or transformed into new musical instruments by the user community without any intellectual property rights (IPR) issues. Users design files to meet their own requirements and, in doing so, upgrade and improve the platform itself. Each platform upgrade enhances the creativity and innovation of the entire user community. In an interview with Kuang (2009), Ziv Bar Ilan—the founder of Zoybar—elaborated on his stance on decentralization in design and the absence of a single authorial vision in Zoybar.

> Cliff Kuang: Product design often works best when it's guided by a single vision. What advantages do you think open-source hardware design has, over that model of design?
> Ziv Bar Ilan: I tend to avoid the use of the term *open-source hardware* since it hasn't been settled yet. I focus on the innovation model and its benefits. Open-hardware architecture is just one aspect of the concept. Decentralized innovation combines a collective vision with independent initiative. Anyone can join and contribute. Centralized R&D labs, however, will always be limited by their selective group of people and their restricted environments. Every independent developer can modify, hack, and attach third-party components according to his unique vision. Zoybar's open hardware does most of the heavy lifting and shortens the development process. Independent developers can now focus entirely on the development of new features that would be relevant to the whole Zoybar community. Users can then interact and learn from the collective intelligence by sharing their ideas and models on the Zoybar community Website. This open process lowers the uncertainties of demand and usability of every new feature with real-time feedback. It provides support and incentives for actual needs rather than dictating unwanted products.
> Excerpt from the interview of Ziv Bar Ilan by Cliff Kuang. (Kuang 2009)

In distributed cognition (Norman 1993), a group of minds has no head or leader. As such, "How can a mind function so effectively and creatively without anyone (or anything) in charge?" (Resnick 1997, xiv). Is the murmuration of birds in the evening sky a grand opera orchestrated by a single leading mind? Resnick (1997) argues that the flock of birds is not a big bird; neither do they follow a leader nor make patterns through chance. Members of the flock, like the continuously changing teams of *Banārasī* toy designers based on the requirements of the client orders, are organized and coordinated without an organiser. Resnick does not reject the relevance of centralized thinking but argues that decentralized phenomena are self-evolving, self-correcting, and sustainable. Decentralized systems enable communities to make decisions at local levels, which together form a cohesive whole in which every participant is an active stakeholder.

In his seminal book *Turtles, Termites, and Traffic Jams* (1997), Resnick discusses several decentralized patterns through his software *StarLogo* and attempts to debunk the myth of a centralized mindset. For instance, the book explains the dynamics of how traffic jams occur. Are they caused by accidents on the road? Do some cars not follow the rules? Does road maintenance disrupt the flow of traffic? Resnick followed a simple model, randomly placing several cars on a road simulation driving at different speeds to different destinations. Interestingly, without any singular cause/event/person like roadblocks, car accidents, or new construction, traffic jams still occurred. While the cars moved forward in the traffic jam as a singular object, the traffic jam as an "object" moved backward. A traffic jam isn't just a collection of cars but an "object" made of individual cars. Such self-organizing systems represent a different type of organization and a novel way of thinking and knowing.

Despite numerous studies exploring decentralized systems, various forms of cooperative and collaborative decision making, and stigmergy, there continues to be a persistent overestimation of the creative capacity of the unaided mind of the individual. In reality, modern design environments are characterized by the highly distributed nature of design teams, the diversity of tools, and the complexity and dynamics of the design environment (Wang et al. 2002, 984 & Craig &, Zimring 2000), necessitating delicate negotiations between team members to reach effective design solutions. Arias et al. (2000, 87) highlight the critical role of externalizations—such as sketches, models, or written descriptions—as follows: they create a tangible record of mental processes, represent artefacts that can be examined by others, and provide the foundation for critique and negotiation among team members. However, collective decision making is impeded by the absence of a single head or central figure in collaborative environments. This limitation is central to the challenge of collaborative design, as noted by both Arias et al. (2000, 86) and Cross and Cross (1995, 144).

If the challenge of collaboration without a single head or central leadership is true, one would expect seamless collaboration between reflective practitioners to be rare. However, research on the *Banārasī* community of toy designers suggests otherwise. These designers manage to collaborate effectively despite several factors that challenge collaborative efforts in formalized structures of modern design: the absence of a clear leader, the prevalence of tacit knowledge, and the fluidity of team membership. Rather than depending on externalizations or formal structures, the *Banārasī* community of toy designers seems to be relying on long-established practices of emulation, shared material consciousness, and ingrained social networks that naturally facilitate knowledge transfer and problem solving.

As discussed in earlier chapters, *Banārasī* toy makers informally form different groups based on shared skills, proximity, and familiarity for every new order. Each toy maker produces only a part of the toy, while the rest of the design decisions, and production are passed on to others in the group, often with few or no forwarding instructions. Surprisingly, successive toy designers are expected to continue shaping the toy and complement the changes without knowing the plans or decisions of their predecessors. However, after the assembly line is completed, the finished toy is coherent and reflects a *Banārasī* identity. The seller and the consumer, often sharing the same context as the producers in the same time and space, can give immediate feedback on the acceptability of these modifications. They are not only part of the financial transaction but also stakeholders in the production of worldview-appropriate and acceptable artifacts. The unwritten but implicitly followed conventions, canons, and beliefs by all stakeholders (toy designers, sellers, and consumers) lead to a self-corrective mechanism that quickly discourages and moderates inappropriate design changes. By itself, this is hardly surprising, but here the "mastery resides not in the master but in the organization of the community of practice of which the master is a part" (Lave and Wenger 1991, 4).

Although the field of design is largely marked by a centralized mindset moderating change through a leader, in recent times, partially leaderless decentralized thinking has been observed in co-design or participatory design methods. Peer-to-peer exchanges of knowledge for transfer, upgrading, and building in the absence of a facilitator, credit, or perks have introduced cooperative thinking not only among designers but also between designers and users. Design decisions are no longer restricted to designers alone but now encompass all stakeholders of the product. This marks a shift of design power from the hands of capitalists through the designer to the hands of stakeholders as users and knowers. The moderation of opinions reflected in design decisions through informal and, at times, formal systems is similar to the "community of practice" described by Wenger (1999). The community of practice consists of the agent, an activity, and the world. The agents or community members share a common concern or passion (activity) for something they do and learn how to do it better as they interact regularly. Like the *Banārasī* community, members of such horizontal practices also learn from a common pool of knowledge of the lived-in world, which this study considers as the worldview. These self-organizing units are defined by knowledge rather than by tasks, where participation is not only a response from a stakeholder but also a need for mutual engagement to solve the problem at hand.

Analogous to the *Banārasī Khilōnā*, Wenger believes that knowledge and ways of perceiving and manipulating objects are encoded in artefacts within a community of practice. It is around artefacts that the practice of doing and approaching things revolves in the community. However, these practices may not necessarily involve multiple members in creating the artefact. The Yucatec midwives, Vai and Gola tailors, naval quartermasters, meat cutters, and the Association of Nondrinking Alcoholics discussed by Lave and Wenger are largely authored or create artefacts through a singular vision governed by the common pool of community knowledge. On the other hand, multiple members of the *Banārasī* practice of toy design co-create a toy by separately conceptualizing its different parts in the absence of oral or written forms of knowledge transfer. This further pushes the boundaries of the theory of community of practice because it extends participation, retention, and change in a worldview-shaped artefact far beyond the centralized mindset of authorship and singular vision that can be widely applied in not just theories of learning but also in the practice of organizations and technology.

The insights generated through the study of the *Banārasī* toy makers are indicative, though not representative, of the vast scope of the lesser explored role of self-organizing tendencies and distributed cognition in collaborative practices. The elicitation of the shared worldview of the *Banārasī* toy designers sheds light on the behaviour of all stakeholders within a decentralized setup for decision making. It is indeed difficult to elicit, construct, or reveal such a shared worldview of all communities, but the study of decentralization may further benefit from the knowledge of conventions, standards, and customs that promote or deter decentralized behaviour among members in the absence of a leader. Such ways of knowing, thinking, and doing can become richer from a mutual exchange of insights and practices, which broadens the framework of competence and experience through the lived-in world.

4 Contemporary Pedagogy

In the summer of 2017, through the mobility funds granted by the Ministry of Education, Finland, for collaboration between Finnish and Indian universities, I had an opportunity to visit a few Finnish early education center. In a second-grade English classroom, several students intently focused on the patterns drawn on the blackboard and then meticulously used the needle and thread to replicate the same on a matte cloth. In pin-drop silence, the class concentrated on the pattern (perhaps their intense concentration was also to prevent hurting themselves with the needle). For forty-five minutes, not a word of English was read, spoken, or written. The English class ended with the teacher collecting the embroidery material of each student and promising to continue the same next week. Later, I asked the teacher how she proposed to teach the English language to students with embroidery. She explained that embroidery and stitching build the fine motor skills of children, which are essential for holding pencils when writing. Therefore, it is a practice in several Finnish schools to introduce embroidery early on to support the writing skills required for language acquisition. This pedagogy resonates with the claims of recent neuroscience studies that there is a direct correlation between fine motor skills, hand–eye coordination, and the development of the brain Björklund L (2009).

In 1865, Uno Cygnaeus introduced a cultural-design-based education system that is still followed in Finland, Norway, Sweden, and Denmark. This learning-by-doing approach is called sloyd education, and research on the pedagogy of cultural design is established under the discipline of *slöjdpedagogik*. *Sloyd* originally meant "handy or skillful" and referred to the making of cultural designs (Þorsteinsson et al. 2015). However, in relation to education, *Sloyd* refers to discussions among philosophers regarding the pedagogical value of cultural design within general education (Borg 2007). According to the Sloyd Training School in Boston, "Sloyd is tool work so arranged and employed as to stimulate and promote vigorous, intelligent self-activity for a purpose, which the worker recognises as good" (Rich 1905, iii). Cultural design continues to be an integral part of Scandinavian education, even in advanced classes where students are expected to learn textiles, stitching, carpentry, and pottery; many upper secondary schools offer a diploma in crafts, too.

An extension of the Finnish Sloyd pedagogy can be observed in the "paper sloyd" for primary classes developed by Ednah Anne Rich as a preparatory skill for progressing to more challenging handwork related to textiles and woodwork in advanced classes. Simply speaking, paper sloyd is origami or paper cultural design. However, in the form of pedagogy, a relationship between numbers and language is established while folding the paper; this relationship is easily applied to formal subjects later. According to Rich, "Observation is quickened; eyes are trained to see right lines and distances, thus aiding in free-hand drawing and writing; while the hand and wrist muscles, being used for a definite purpose, unconsciously become obedient assistants" (Rich 1905, 1).

The hand and mind dexterity developed in paper folding is applied to more complex forms of cultural design like textile and carpentry in advanced classes.

From simple problems like measuring a line, students are exposed to complex calculations of a warp and weft on a loom. It is not uncommon to see Finnish students use trigonometry to figure out the angles of complex carpentry assignments in wood.

The Indian equivalent of sloyd education is M. K. Gandhi's *Nayī Ta'līm*. Similar to Finland, *Nayī Ta'līm* was also envisioned as a part of nation-building. The simplest handicraft suitable for all, including gender, and socially and economically relevant to the whole of India, was undoubtedly spinning for M. K. Gandhi. This labour-intensive process, if introduced in educational institutions, should fulfill three purposes: make education self-supporting, train the bodies of the children as well as their minds, and pave the way for a complete boycott of foreign yarn and cloth in a colonized land. The children, thus equipped, would become self-reliant and independent. Gandhi suggested that handicrafts should be taught to school children to develop their intellect and increase their productivity in early life. In his own words, "Surely, if the State takes charge of the children between seven and 14 and trains their bodies and minds through productive labour, the public schools must be frauds and teachers idiots if they cannot become self-supporting" (1963, 256).

Gandhi astutely perceived that unless one knows arithmetic, one cannot report how many yards of yarn have been produced on the *Taklī*—an Indian spindle used in the *Charkhā* or a spinning wheel—how many standard rounds it will make, or what the count of the yarn spun is. For this, knowledge of figures, knowledge of operational mathematics are a must. Take geometry, for instance. What can be a better demonstration of a circle than the disc of the *Taklī*? One can teach all about the circle with a *Taklī* without even mentioning Euclid.

The materials also provoke a scientific enquiry. Why is the *Taklī* made of a brass disc and uses steel in its spindle? The more primitive *Taklīs* consisted of a wooden spindle with a disc of slate or clay. What could be the reason for shifting the design from wood and slate to brass and steel? Why does the disc have that particular diameter, no more and no less? It is the job of the teacher to ignite the curiosity of the student toward such a scientific frame of mind, according to Gandhi, with the help of the technology and artefacts of weaving. This is a constructionist belief that knowledge is constructed when learners engage in the hands-on creation of and sharing of physical artefacts (Papert 1991).

In today's world, Gandhi's *Nayī Ta'līm* may appear to be an obsolete. But it envisioned the radically creative curriculum of advanced countries like Finland that upholds cultural design and making as key to education. Can we reconnect with cultural design for a transformative paradigm shift in education?

Mjelde (1997 & 2006), in her argument for a "workshop pedagogy", advocates a learning curve from practice to theory, as well as a much desired, equitable conversation between the head and the hand education:

> One of the main challenges created by Reform 94 is, how can we possibly link in the general subjects with workshop learning and vocational theory. Today, students learn the theorems of Pythagoras as geometry, as general theory, rather than, for example, in the context of how a carpenter uses Pythagoras in building a house. When carpenters calculate an angle, they are applying Pythagorean theorems. Why can they not learn Pythagoras as they learn to calculate or measure an angle with the tools of their intended trade? Printers use chemicals everyday.

> Why could a teacher of the natural sciences not work together in the same place with the workshop teacher? Why could s/he not teach about the chemicals that one uses in the graphics field on an everyday basis? Student cooks learn the principles and formulas of the functioning of flame, steam, and biochemistry when they are in the academic classroom, while they actually experience the application of this knowledge, albeit unconsciously, in practice in the kitchen. Ratios of temperature and selection and quantities of ingredients learned in the kitchen seem distant from the principles of biochemistry learned in the classroom. Why could the natural science teacher not work together with the teacher of cooking to teach the principles and formulas in the kitchen? (Mjelde 2006, 82–83)

In India, the disjunction between skill and theoretical discourses has become so widespread that it now looms largely in the masses of graduate and uneducated youth who lack employable skills. The skilled cultural designer is deemed uneducated, and the formally educated person lacks employable skills! Cultural design as the transformative bridge that makes education useful never found its place in Indian education—notwithstanding Gandhi. As a result, there is a pedagogical crisis in India that pervades all educational streams; without real-world applications, concepts remain alien and thus the foundation of innovative thinking is whittled away such that no amount of government-funded schemes for innovation can restore it. Unfortunately, cultural design apprenticeship as a mode of transmission acquisition is no longer deemed viable in contemporary work organizations, where continuous change is the norm (Guile and Young 1998; Sennett 1998). The failure to integrate cultural design into mainstream education has resulted in an imbalance in that practical, employable skills are rare among the educated, while those with practical skills are viewed as uneducated and cut off from opportunities for social mobility.

Similarly, the literature on formal design curriculum often discusses the divide between the hand and the mind but resigns itself to the comfort of the established mind-dominated pedagogy. Nevertheless, in many countries, a new form of apprenticeship is being revived as a way of encouraging employers to provide access to unemployed youth who need to make the transition from school to work. In South Africa, the term *apprenticeship* has been replaced by *learnership* to signal a break with practices of the past seen as elitist and racist. The Igbo apprenticeship model followed by the Igbo tribe in Nigeria after the Biafra Civil War is claimed to have drawn the majority of Igbo families out of poverty. It is widely recognized as the backbone of the Nigerian economy (Orugun and Nafiu 2014) today. In this model, established businesspeople accept apprentices and train them in their trade to support their own businesses. As the apprentices build their capacity to contribute to the profits of their mentor, they also gain the experience to handle a company on their own. When the businesspeople deem the apprentices ready, they offer them startup capital to set up their own businesses (Iwara et al. 2019; Kanu 2017).

The Nordic model of apprenticeship is recognized for its seamless integration of vocational education into the school system and its continuation through hands-on, practice-based internships in collaboration with industry during higher education. This approach emphasizes early exposure to practical skills and thus bridges theoretical learning with real-world application, a process that contributes significantly to the employability of students.

A key feature of the Nordic model is the close collaboration between social partners, trade unions, and a committed welfare state policy that supports vocational education and apprenticeships. These institutional frameworks ensure that the apprenticeship systems remain effective and beneficial for both students and industries.

However, there are notable differences in how the apprenticeship systems are implemented across Denmark, Sweden, and Norway, despite being grouped under the Nordic model. According to Nyen and Tønder (2020), Denmark has a dual apprenticeship-based vocational education model, which differs from Sweden's school-based model. During hands-on industrial training, Swedish students retain their student status, whereas Danish students are employed as apprentices. Norway has developed a hybrid model, with apprenticeship training following school-based education in the 2 + 2 model.

In the East, the Japanese model of apprenticeships—learning by doing—continues to be studied for its efficacy and cultural distinctiveness in management circles. The traditional apprenticeship system, originating in the cultural design communities of Japan has heavily influenced the management pedagogy. The persistent practice of the foundational elements of the discipline—calligraphic stroke (*Kanji*), sake glass (*guinomi*), or the constant repetition of "Twinkle, twinkle little star" in Suzuki violin classes, all of which are instances of observation and emulation—translates into the management domain as "continuous learning." Singleton calls such patterns of learning an "introduction to a Japanese 'social technology'—a body of cultural practice used to achieve social learning purposes" (1998, 17). There is no didactic tell-and-teach method of transfer of knowledge. In fact, given the deep understanding and respect of tacit knowledge, the apprentice system strongly adheres to the principles of learning by constant practice. A vivid example of this constant awareness of being a learner is the efforts of Matsushita Electric to build a bread-making machine, as discussed by Nonaka and Takeuchi (1995) in their seminal work on how Japanese companies innovate.

The success of the Toyota Production System is attributed to the Kaizen system, which was borrowed from the traditional apprentice model of Japan. Magnier-Watanabe (2011) defines Kaizen as a systematic approach to recognizing problems in current practices and working in inclusive teams to devise and immediately implement ways to improve the process. These solutions are a set of ongoing improvements or improvisations and not an annual overhaul.

Any improvement arrived at by consensus is subsequently standardized before further improvisations are proposed (Imai 1986). Therefore, according to Storz (2008), Japanese management invests in the development of firm-specific human capital at the point of entry through apprenticeship, job rotation, and cross-functional training to prepare their staff to continually improvise. Such apprenticeship gives the advantage of developing firm-specific knowledge, disseminating it in the organization at all levels, and immediately embodying it in new technologies in what Nonaka and Takeuchi (1995) call the 'knowledge-creating' companies.

5 Cultural Maker Labs

The *Kārkhānā*, also called the design studio, has been a neglected subject of critical inquiry (Hoffmann 2012, 12), so much so that post-modern discourse has written its obituary in the post-studio episteme of making and displaying. A parallel discourse persists that celebrates the vitality of the studio and recognises the *Kārkhānās* as the predecessors of the maker movement itself. The stereotypical assumptions that maker spaces are contemporary enclaves of advanced technologies, dominated by white males and driven by the global West, have been increasingly challenged by scholars. Epistemological antecedents position the maker culture as a grassroots movement characterized by low-tech, handmade, and bricolage practices that shape what is made, how it is made, how it is learned, and for/by whom within a local context. The *kārkhānās*, or cultural maker labs, have long served and continue to serve as spaces for individual experimentation, open sources for emulation, repositories of low-tech tools and technologies, and equitable organizations for dividing tasks into meaningful work.

Hatch (2014), in the *Maker Manifesto*, identified nine fundamentals of the maker movement: make, share, give, learn, tool up, play, participate, support, and change (p. 11). These principles can be traced back to the erstwhile maker labs—*Kārkhānās*. *Kārkhānās* are decentralized, low-tech maker labs that develop organically within the local ecosystem of technologies, tools, raw materials, and skills to meet the needs of the community. They serve as spaces for living, learning, and earning a livelihood for their owners, who not only possess the means of production but also become open sources of know-how. These spaces drive experimentation to meet demand, allow owners to reap profits and bear losses, and function as alternative nano-entrepreneurial models. The conventional Marxist distinction between the bourgeois—owners of capital—and the proletariat, who serve the owners, often does not apply in these contexts. Here, the proletariat own the means of production, though they may not necessarily adhere to a capitalist profit model. This invites us to consider the significant points of departure between the informally organized *Kārkhānās* as maker labs and the consciously and formally designed high-tech maker labs of today that focus on personal innovation.

Martin (2015) cautions against the seductive yet fundamentally flawed conceptualization of the maker movement, which assumes that its power lies primarily in its high-tech toolset and that these tools are the catalysts for transformation in education. Maker spaces come in various shapes and sizes and serve as gathering points for tools, projects, mentors, and expertise. Yet it is not merely the collection of tools but the potential to create that defines a maker space (Hlubinka et al. 2013). Low-tech maker labs do exist and function with a capacity for risk taking, improvisation, and innovation. The difference between technologies may not rest on the binary of high-tech versus low-tech, but rather on the appropriateness, affordability, and acceptability of tools within each context, and how they fit into the learner's developmental trajectory.

It is also crucial to examine the role of technology in a low-tech maker lab dominated by generational knowledge. In *Kārkhānās*, technology and tools are employed to assist the human hand, not to replace it. The tool's capability is not what is advanced or cutting edge; rather, it is the designer's advanced ability to continuously push human limits to craft a form that closely mirrors the *eidos*. In cultural design, tools are intended to overcome the limitations of the human body—where the body ends, the tool begins, both figuratively and physically. In contrast, mechanization is not about extending the human body's capabilities but about reducing the effort required by the human body or displacing humans from the process of making.

Gershenfeld (2005) emphasizes that maker spaces are not only places of production but also pedagogical environments where ordinary folks can solve their problems by creating the tools they need rather than purchasing or outsourcing them. This perspective has spurred considerable scholarship on the pedagogical potential of high-tech contemporary maker labs. However, questions remain about whether these environments genuinely foster learning given their strong focus on acquiring technology rather than knowledge. In contrast, *Kārkhānās* are inherently educational spaces that follow a master–apprentice model of work, a structure notably absent in modern maker labs.

Learning in *Kārkhānās* progresses through hands-on construction and observation. Apprentices gain understanding by either creating objects themselves or witnessing the construction process led by a more advanced peer or master. This embedded social structure naturally facilitates learning, in contrast to the often isolated, self-directed learning that occurs in modern maker labs. In contemporary maker labs, knowledge can be "sticky," meaning it is difficult to transfer or share with others. In contrast, cultural design communities employ emulation as a primary method for learning, upgrading, and preserving knowledge. This practice ensures that know-how becomes the collective property and legacy of the community and is passed down through generations.

Moreover, while modern maker labs are typically designed to create artificial environments that are disconnected from the lived reality of the makers, cultural designers often live in their *Kārkhānā*, which is simply an extension of their home. These maker labs are living laboratories of tradition as opposed to artificially constructed spaces intended for invention and innovation through technology. In *Kārkhānās*, the focus is not merely on access to tools and technology but on the opportunity to learn how to use them for personal or socially meaningful purposes. This approach is vital for fostering innovation and learning among masters and apprentices.

Marchand (2016) explains that becoming an expert cultural designer is synonymous with being aware of the available resources in the *Kārkhānā* and knowing how to manipulate and employ them competently to solve problems. Grassroots makers naturally possess the motivation, need, and skill to innovate, often through tacitly absorbed situated learning rather than formal instruction. This ability to innovate under resource constraints is exemplified by practices like Indian *Jugaad*, Chinese *Shenzai*, Brazilian *Jeitinho*, and Japanese *Kaizen*, which all emphasize improvisation across cultures. These ecosystems have shown remarkable resilience through political and economic upheavals and continually adapt to meet new challenges. Masters in these systems have not only adapted their skill sets, materials, and technologies but also their business strategies, and they pass on the ability to innovate, adjust, and market to meet new demands.

Therefore, there is a need to acknowledge *Kārkhānās* as multimodal, open-innovation maker labs that operate locally and integrate the concurrent processes of living, learning, and generating livelihood. These spaces accommodate real-life situations involving various community stakeholders, including the designers themselves, and should be recognized for their holistic approach to innovation and education.

In conclusion, this text proposes ways of knowing, thinking, and doing that can enrich the practice of modern design and also be enriched by a mutual exchange of insights and practices, thereby broadening the framework of competence and experience for both.

References

About EPCH—Promoting and Facilitating the Export of Indian Handicrafts Across the Globe (n.d.) https://www.epch.in/about-epch

Al-Dajani H, Marlow S (2013) Empowerment and entrepreneurship: a theoretical framework. Int J Entrep Behav Res 19(5):503–524

Alexander C (1964) Notes on the synthesis of form. Harvard University Press

Archer M (1959) Tippoo's tiger—museum monograph no. 10. HM Stationery Office, London

Arias E, Eden H, Fischer G, Gorman A, Scharff E (2000) Transcending the individual human mind—creating shared understanding through collaborative design. ACM Trans Comput-Hum Interact (TOCHI) 7(1):84–113

Bell E, Mangia G, Taylor S, Toraldo ML (2018) The organization of craft work

Björklund L (2009) The forming and assessment of creative skills, from a neurocognitive point of view. In: PATT-22 conference, Delft, The Netherlands, 24–28

Blundel RK, Smith DJ (2013) Reinventing artisanal knowledge and practice: a critical review of innovation in a craft-based industry. Prometheus 31(1):55–73

Bofylatos S (2017) Adopting a craft approach in the context of social innovation. Craft Res 8(2):223–240

Borg K (2007) Processes or/and products–what do teachers assess? Des Technol Educ Int J 12(2):57–65

Brinck I (2024) Craft thinking: a relational approach to making and design. In: Craft and design practice from an embodied perspective. Routledge, pp 30–39

Buchanan R (1992) Wicked problems in design thinking. Des Issues 8(2):5–21

Bunn S (2022) Creative movements: hands, arms, materials and words in making baskets. In: Knowing from the inside: cross-disciplinary experiments with matters of pedagogy, p 81

Cambourne B (2022) Why a teacher's beliefs matter: using a theory of learning to explore instructional decisions. J Read Recov 22(1):53–69

Collier AF (2011) The well-being of women who create with textiles: implications for art therapy. Art Ther 28(3):104–112

Colman A (2006) A dictionary of psychology. Oxford University Press, New York, p 670

Craig DL, Zimring C (2000) Supporting collaborative design groups as design communities. Des Stud 21(2):187–204

Cross N, Cross AC (1995) Observations of teamwork and social processes in design. Des Stud 16(2):143–170

Dorst K (2011) The core of 'design thinking' and its application. Des Stud 32(6):521–532

Duhigg C (2012) The power of habit: why we do what we do in life and business, vol 34, no 10. Random House
Ferris M (2009) Editor's introduction. In: Making futures: the crafts in the context of emerging global sustainability agendas. University of Plymouth, Devon, UK, pp i–vi
Friedman TL (2005) It's a flat world, after all. New York Times 3:33–37
Gandhi M (1963) Village swaraj. Narajivan Publishing House
Gershenfeld NA (2005) FAB: the coming revolution on your desktop—from personal computers to personal fabrication. Basic Books (AZ)
Gibson M (2019) Crafting communities of practice: the relationship between making and learning. Int J Technol des Educ 29:25–35
Gladwell M (2006) The tipping point: how little things can make a big difference. Little, Brown
Groth C, Mäkelä M, Seitamaa-Hakkarainen P (2013) Making sense. What can we learn from experts of tactile knowledge? FormAkademisk 6(2)
Guile D, Young M (1998) Apprenticeship as a conceptual basis for a social theory of learning. J Vocat Educ Train 50(2):173–193
Hatch M (2014) The maker movement manifesto. McGraw Hill Professional
Hlubinka M, Dougherty D, Thomas P, Chang S, Hoefer S, Alexander I, McGuire D (2013) Makerspace playbook: school edition. Retrieved from Maker Media website: http://makerspace.com/wpcontent/uploads/2013/02/MakerspacePlaybook-Feb2013.pdf
Hoffmann J (2012) The studio. The MIT Press, Cambridge
Huotilainen M, Rankanen M, Groth C, Seitamaa-Hakkarainen P, Mäkelä M (2018) Why our brains love arts and crafts: implications of creative practices on psychophysical well-being. FormAkademisk 11(2). https://doi.org/10.7577/formakademisk.1908
Imai M (1986) Kaizen, vol 201. Random House Business Division, New York
Iwara IO, Amaechi KE, Netshandama V (2019) The Igba-boi apprenticeship approach: arsenal behind growing success of Igbo entrepreneurs in Nigeria. Afr J Peace Conflict Stud 227–250
Jaatinen J, Lindfors E (2019) Makerspaces for pedagogical innovation processes: how Finnish comprehensive schools create space for makers. Des Technol Educ 24(2):n2
Jakob D (2013) Crafting your way out of the recession? New craft entrepreneurs and the global economic downturn. Camb J Reg Econ Soc 6(1):127–140
Jones JC (1970) Design methods: seeds of human futures. Wiley, New York
Kanu IA (2017) Igwebuike and the logic (Nka) of African philosophy. IGWEBUIKE Afr J Arts Humanit 3(1):1–13
Kuang C (2009) We build the parts. You build the product. Accessed 20 Mar 2013. http://www.fastcompany.com/1293079/we-build-parts-you-build-product
Lave J, Wenger E (1991) Situated learning. Legitimate peripheral participation. University of Cambridge Press, Cambridge
Law J (2004) After method: mess in social science research. Routledge
Luckman S, Thomas N (2018) Crafting economies. Bloomsbury Publishers
Lury C (2012) Inventive methods: the happening of the social. In: Wakeford N (ed). Routledge, London
Magnier-Watanabe R (2011) Getting ready for Kaizen: organizational and knowledge management enablers. Vine 41(4):428–448
Migueles JH, Cadenas-Sanchez C, Esteban-Cornejo I, Torres-Lopez LV, Aadland E, Chastin SF, Erickson KI, Catena A, Ortega FB (2020) Associations of objectively-assessed physical activity and sedentary time with hippocampal gray matter volume in children with overweight/obesity. J Clin Med 9(4):1080
Maidment J, Macfarlane S (2013) Older women and craft: extending educational horizons in considering wellbeing. In: Promoting health and well-being in social work education. Routledge, pp 145–156
Marchand THJ (ed) (2016) Craftwork as problem solving: ethnographic studies of design and making. Ashgate Publishing, Ltd.

Martin L (2015) The promise of the maker movement for education. J Pre-Coll Eng Educ Res (J-PEER) 5(1):4
McIntyre MH (2010) Consuming craft: the contemporary craft market in a changing economy. Craft Council, London
Mjelde L (1997) The promise of alternative pedagogies: the case of workshop learning. In: Educational dilemmas: debate and diversity. Cassell, London
Mjelde L (2006) The magical properties of workshop learning, vol 2. Peter Lang
Moore G (2005) Humanizing business: a modern virtue ethics approach. Bus Ethics Q 15(2):237–255
Nelson HG, Stolterman E (2014) The design way: intentional change in an unpredictable world. MIT Press
Nonaka I, Takeuchi H (1995) The knowledge creating company. Oxford University Press, New York
Norman D (1993) Things that make us smart: defending human attributes in the age of the machine. Addison-Wesley Longman Publ. Co.
Nyen T, Tønder AH (2020) Capacity for reform: the changing roles of apprenticeship in three Nordic countries. Transfer Eur Rev Lab Res 26(1):43–56
Orugun JJ, Nafiu AT (2014) An exploratory study of Igbo entrepreneurial activity and business success in Nigeria as the panacea for economic growth and development. Int J Sci Technol Res 3(9):158–165
Papert S, Harel I (1991) Situating constructionism. Constructionism 36(2):1–11
Patil K, Athavankar U (2022) Design precedents to design innovation: category-based reasoning in problem-solving. Des J 25(1):62–85
Pöllänen S (2015) Elements of crafts that enhance well-being: textile craft makers' descriptions of their leisure activity. J Leis Res 47(1):58–78
Þorsteinsson G, Olafsson B, Yokoyama E (2015) Otto Salomon and Aksel Mikkelsen, and their pedagogical models for establishing sloyd education in Denmark and Sweden
Reitan JB (2014) Learning by watching vernacular Iñupiaq-Inuit design learning as inspiration for design education. Techn Ser A 21(2):1–20
Resnick M (1997) Turtles, termites, and traffic jams. MIT Press, London
Reynolds F (2000) Managing depression through needlecraft creative activities: a qualitative study. Arts Psychother 27(2):107–114
Rich EA (1905) Paper sloyd: a handbook for primary grades. Ginn
Rittel HWJ, Webber MM (1973) Dilemmas in a general theory of planning. Policy Sci 4(2):155–169
Sennett R (2008) The craftsman. Yale University Press, New Haven
Sforzi J, Colombo LA (2020) New opportunities for work integration in rural areas: the 'social flavour' of craft beer in Italy. Sustainability 12(16):6351
Sheridan K, Halverson ER, Litts B, Brahms L, Jacobs-Priebe L, Owens T (2014) Learning in the making: a comparative case study of three makerspaces. Harv Educ Rev 84(4):505–531
Silliman SW (2009) Change and continuity, practice and memory: native American persistence in colonial New England. Am Antiq 74(2):211–230
Simon HA (1969) The sciences of the artificial. MIT Press, Cambridge, MA
Singleton J (ed) (1998) Learning in likely places: varieties of apprenticeship in Japan. Cambridge University Press
Storz C (2008) Dynamics in innovation systems: evidence from Japan's game software industry. Res Policy 37(9):1480–1491
Stronge S (2009) Tipu's tigers. V&A Publishing, London
Teiwes H (1996) Hopi basket weaving: artistry in natural fibers. University of Arizona Press
Wang L, Shen W, Xie H, Neelamkavil J, Pardasani A (2002) Collaborative conceptual design—state of the art and future trends. Comput Aided des 34(13):981–996
Wenger E (1999) Communities of practice: learning, meaning, and identity. Cambridge University Press

Glossary

1	Āratī	आरती	A ceremony performed in worshipping a God: a dish holding a lamp, burning ghee, incense or other articles is moved in a series of circles in front of the idol (Mcgregor 1997: 92)
2	Abhinav	अभिनव	Quite new, novel, recent (Bahri 2009: 107)
3	Acakan	अचकन	A tight-fitting long coat (Bahri 2009: 32)
4	Adhārmik	अधार्मिक	1. Unjust, unrighteous; wicked: sinful 2. Irreligious (Mcgregor 1997: 25)
5	Almārī	अलमारी	1. Cupboard 2. Wardrobe 3. Bookcase (Mcgregor 1997: 58)
6	Ālthā	आलथा	Lac dye, a kind of red dye (used by Hindu women for staining their feet red) (Bahri 2009: 125)
7	Andāz	अंदाज़	(1) Rough estimate, guess, conjecture (2) Mode, manner, style (Bahri 2009: 14)
	Apsarā	अप्सरा	(1) Nymph, a female dancer in the court of Indra (2) Fairy (Bahri 2009: 102)
8	Asabhya	असभ्य	Uncivilized, uncivil, unrefined, rude (Bahri 2009: 143)
9	Aṣṭādhyāyī	अष्टाध्यायी	Sanskrit treatise on grammar written in the sixth to fifth century BCE by the Indian grammarian Panini. This work set the linguistic standards for Classical Sanskrit. It sums up in 4000 sutras the science of phonetics and grammar that had evolved in the Vedic religion. Panini divided his work into eight chapters, each of which is further divided into quarter chapters (Encyclopædia Britannica Inc. 2014)
10	Aṣṭadhātu	अष्टधातु	Eight metals (gold, silver, copper, lead, tin, zinc, iron and mercury) (Bahri 2009: 141)
11	Aśoka	अशोका	The tree *Jonesia asoka* known especially for its red flowers (Mcgregor 1997: 66)
12	Astra Pūjā	अस्त्र पूजा	Astra-Missil, weapon (Bahri 2009: 147) Pooja-to worship (Mcgregor 1997: 634) Astra Pooja-worshiping weapons

K. Patil, *Toying with the Mind*, Design Science and Innovation,
https://doi.org/10.1007/978-981-96-9227-9

13	Babbar Śer	बब्बर शेर	Lion (Bahri 2009: 1193)
14	Bābul	बाबुल	Father (Bahri 2009: 1213)
15	Bacce/Bacchoṅ	बच्चे/बच्चों	Children, baby, kids (1183)
16	Bagīcā	बगीचा	A small garden, a flower garden (Bahri 2009: 1182)
17	Baiṇḍ-bājā	बैंड-बाजा	Colloquial term for local band or orchestra. Baja refers to a musical instrument
18	*Bāṇārasīpan*	बनारासिपान	*Bāṇārasī* term for *Bāṇārasīness* or *Bāṇārasī* identity
19	Bhagavat	भगवत	Having to do with Bhagwat or Vishnu Pooja (Mcgregor 1997: 763)
20	Bhagavat Pāṭha	भगवत पाठ	1. Reading; study especially of a sacred text 2. Recitation of a sacred text (Mcgregor 1997: 620)
21	Bharāī	भराइ	Name given to a token gift of a coconut, rice, and c. made to a bride (Mcgregor 1997: 277)
22	Bhela	भेल	A mixture; sthg. Mixed in (Mcgregor 1997: 773)
23	Bindī	बिंदी	Dot, mark or spangle ornamenting the forehead (Mcgregor 1997: 729)
24	Brahmā	ब्रह्मा	All pervading spirit of the universe (Bahri 2009: 1243)
25	Brahmavaivarta Purāṇas	ब्रह्मव्रत पुराण	Purāṇas—a class of voluminous work in Sanskrit dealing with aspects of ancient Indian history, legend, mythology or theology (Mcgregor 1997: 637)
26	**Chāṭ**	चाट	An appetizer, savory; a delicacy as a spiced fruit or vegetable salad (Mcgregor 1997: 310)
27	Cādar	चादर	1. Sheet (of cloth) 2. Bed-sheet (Bahri 2009: 545)
28	Cālū	चालू	Current, running (Bahri 2009: 548)
29	**Cālū Grāhak**	चालू ग्राहक	One who takes, receives or values something; a customer or subscriber (Mcgregor 1997: 282) In this case it refers to a regular customer
30	Chakra	चक्र	1. Wheel 2. Circular object, disc, discus (Bahri 2009: 523)
31	Caṭāk	चटाक	Colloquial term for very bright, highly contrasting or loud
32	Chaṭṭhī	छटी	Sixth day of the Lunar fortnight (Mcgregor 1997: 337)
33	Citrakār	चित्रकार	Painter, artist (Bahri 2009: 555)
34	Cugtī Ciṛiyā	चुगती चिड़िया	*Bāṇārasī* name for pecking toy
35	Cunrī	चुनरी	Partly-colored over-cloth worn by woman particularly girls (Bahri 2009: 564)
36	Cūsnī	चुसनी	Sucking nipple (Bahri 2009: 567)
37	Dānewālī Ciṛiyā	दानेवाली चिड़िया	*Bāṇārasī* name for pecking toy
38	Dhārī Vālā Śer	धारी वाला शेर	*Bāṇārasī* name for tiger toys
39	Dhārmik	धार्मिक	1. Religious 2. Pious, virtuous, just, righteous (Bahri 2009: 885)
40	Dhotī	धोती	Loin-cloth, a sheet of cloth worn around the waist passing between the legs and tucked in behind (Bahri 2009: 892)
41	**Dukhī Parivār**	दुखी परिवार	*Bāṇārasī* name for a nested doll

42	Durgā	दुर्गा	The goddess Durga, the wife of Shiva (Bahri 2009: 840)
43	Dvārpāl	द्वारपाल	Doorkeeper, gatekeeper, commissionaire (Bahri 2009: 867)
44	Gaddī Vālā Hāthī	गद्दी वाला हाथी	*Bāṇārasī* name for a royal elephant in toys
45	Gaṛhnā	गढ़ना	1. To fabricate 2. To hammer 3. To manufacture (Bahri 2009: 446)
46	Gamchā	गमछा	Thin towel (Bahri 2009: 451)
47	Garuḍa	गरुडा	A large eagle-like bird (Bahri 2009: 454) (a supernatural eagle-like being that serves as Vishnu's mount)
48	Genda	गेंद	Ball (Bahri 2009: 486)
49	**Ghāṭ**	घाट	1. Quay, jetty, landing place on a river bank, ferry 2. Port, harbour (Bahri 2009)
50	Ghumātū	घुमातू	*Bāṇārasī* word for a spinning top
51	Gillī Daṇḍā	गिल्ली डंडा	The game of tipcat (482)
52	Gopī	गोपी	Wife of a milkman, milkmaid (Bahri 2009: 488)
53	**Govardhana**	गोवर्धन	1. Govardhan hill (Bahri 2009: 489) 2. Hill near Vrindavan (Bahri 2009: 493)
54	Guḍḍā	गुड्डा	Male doll (Bahri 2009: 475)
55	Gūlara	गूलर	Wild fig, sycamore, *Ficus glomerata* (Bahri 2009: 484)
56	**Gwāla**	ग्वाल	Cowherd, milkman (Bahri 2009: 498)
57	Gwālan	ग्वालन	A woman of the cow-herd class, milkmaid, milk woman (Bahri 2009: 498)
58	Hāṭ	हाट	Periodic market (Bahri 2009: 1747)
59	**Halāhala**—poison during Samudramanthan	हलाहल—विष समुद्रमंथन	Deadly poison (Bahri 2009: 1741)
60	*Hilāntā*	हिलानता	*Bāṇārasī* word for a pecking toy consisting of a single animal with a two part movement
61	Holī	होली	A Hindu festival of colours (Bahri 2009: 1776)
62	Hum	हम	We (Bahri 2009: 1736)
63	Humhiye	हुम्हिये	*Bāṇārasī* word for 'we'
64	Jātaka	जातक	Tales of the various births of lord Buddha (Mcgregor 1997: 633)
65	Jhuñjhunā	झुनझुना	A ratting toy, rattle (Bahri 2009: 675)
66	Janmāṣṭamī	जन्माष्टमी	Birthday of lord Kṛṣṇa which falls on the eighth day of the dark fortnight in Bhadon (July–August) (Mcgregor 1997: 615)
67	Jugāṛ	जुगाड़	1. Device 2. Arrangement (Bahri 2009: 650)
68	Jaṅglī Hāthī	जंगली हाथी	*Bāṇārasī* name for a wild elephant in toys
69	Kāṭanā	काटना	To cut, to chop, to disconnect, to bite, to intersect (Bahri 2009: 334)
70	Kadam tree	कदम का पेड़	*Nauclea cadamba* tree and its bell-shaped flower (Mcgregor 1997: 298)
71	Kāliya Mardana	कालिया मर्दन	*Bāṇārasī* festival based on the story of Kṛṣṇa killing the serpent—Kalia

72	Kamalnayanī	कमलनयनी	Women having eyes like a lotus, lotus eyed (Bahri 2009: 307)
73	Kārāgāra	कारागार	Prison, jail, goal (Bahri 2009: 343)
74	Karakhanās	कारखाना	Workshop, factory, mill (Bahri 2009: 342)
75	Kartab	करतब	Art, skill, wonderful performance (Bahri 2009: 310)
76	Kasērā	कसेरा	Bronze smith, brazier, pewterer (Bahri 2009: 328)
77	Kāśī Naresh	कशी नरेश	*Bāṇārasī* name for the King of Kashi
78	Khaṇḍita	खंडित	1. Broken into pieces; destroyed; spilt, divided; interrupted, broken off 2. Made imperfect; refuted; contradicted; impugned; disappointed (Mcgregor 1997)
79	Kharāda	खराद	A lathe, to be put on the lathe; fig to be polished, or improved (Mcgregor 1997: 230)
80	Khudāī	खुदाई	Digging excavating; engraving (Mcgregor 1997: 240)
81	Krīḍā	क्रीडा	Play, sport, game, frolic (Bahri 2009: 392)
82	Kumbhāra	कुम्हार	Potter (Bahri 2009: 366)
83	**Kurtā**	कुर्ता	Loose collarless shirt (Bahri 2009: 368)
84	Laṅkā	लंका	Ceylon (now Sri Lanka) (Mcgregor 1997: 1410)
85	Laṭṭū	लट्टू	Spinning top, a plummet, knob, electric bulb (Mcgregor 1997: 1420)
86	Līlā	लीला	Supportive display, God's way or doing (Bahri 2009: 1437)
87	Lomaṛī	लोमड़ी	Fox (Bahri 2009: 1446)
88	Māgha	माघ	The eleventh month of Vikrami calendar (Bahri 2009: 1305)
89	Mahābhārata	महाभारत	A great war epic (Bahri 2009: 1302)
90	Mehmān Ghar	मेहमान घर	Guest chamber, guest house (Bahri 2009: 1349)
91	Melā	मेला	Fair, festival, crowd (Bahri 2009: 1348)
92	Nāgnāthiyā	नाग्नाथैया	*Bāṇārasī* toy based on the fight of Kṛṣṇa with Serpent Kaliya. After the fight, Kṛṣṇa danced on the serpent's head
93	Nāgapañcamī	नागपंचमी	Hindu festival in which snake is worshipped as a deity (Bahri 2009: 921)
94	Nakāśī	नकाशी	Engraving, sculpture, draughtsman ship, carving (Bahri 2009: 901–902)
95	Nakal	नक़ल	Copy, duplicate (Bahri 2009: 901)
96	Nakala Karnā/ Utaranā	नक़ल करना/ उतरना	To copy, to trace, to mimic (Bahri 2009: 901)
97	Nandī	नंदी	Bull of Shiva (Bahri 2009: 899)
98	Nauṭaṅkī	नौटंकी	Folk-drama with music and dance (Bahri 2009: 989)
99	Nayāpan	नयापन	Newness, novelty, modernism (Bahri 2009: 911)
100	Nīlakaṇṭha	नीलकंठ	Blue necked, an epithet of Shiva (Bahri 2009: 981)
101	Pāṇḍava	पांडव	1. Having to do with the Pandavas 2. A descendant (esp. one of the five sons) of Pandu (Mcgregor 1997: 617)
102	Paṇḍita	पंडित	Scholar, learned person, Brahman, priest in a temple (Bahri 2009: 998)
103	Pallū	पल्लू	Border, hem, end of a woman's headwear (Bahri 2009: 1046)

104	Pañcāla Naresh	पंचाल नरेश	1. Having to do with the Panchals 2. A Panchal king or Prince (Mcgregor 1997: 617)
105	Patra	पत्र	Leaf, wing (of a bird), letter, note (1012)
106	Pāyala	पायल	An ornament for the ankles, anklet (Bahri 2009: 1061)
107	Pūjanīya	पूजनीय	Revered, adorable, venerable, highly respectable (Bahri 2009: 1091)
108	Prajāpati	प्रजापति	Lord of creatures: the creator; a title of Brahma, and of other creatures; a king (Mcgregor 1997: 654)
109	Pravacana	प्रवचन	Sermon, homily, discourse (Bahri 2009: 1141)
110	Purāṇa-s	पुरानस	1. Old; ancient; primeval 2. A class of voluminous work in Sanskrit dealing with aspects of ancient Indian history, legend, mythology or theology (Mcgregor 1997: 637)
111	Purohita	पुरोहित	Priest, pastor (Bahri 2009: 1087)
112	Pūṭanāvadha	पुतानावध	*Bāṇārasī* toy representing the killing of Putana at the hands of baby Kṛṣṇa
113	Rādhā	राधा	Name of the chief herd girl or Gopi of Braj, the favourite mistress (or the wife of Kṛṣṇa) (Mcgregor 1997: 861)
114	Rādhā-Kṛṣṇa	राधा-कृष्णा	A slogan of devotees of Kṛṣṇa and Radha (Mcgregor 1997: 861)
115	Rākṣasa	राक्षस	Monster, demon (Bahri 2009: 1386)
116	Rāma	राम	Omnipresent God (Bahri 2009: 1392)
117	Rāmāyaṇa	रामायण	The epic describing the life of Rama (Bahri 2009: 1392)
118	Rāmlīlā	रामलीला	A show presenting episodes from Rama's life (Bahri 2009: 1392)
119	Rāslīlā	रासलीला	Kṛṣṇa dalliance and dance with the herd girls of Braj (Bahri 2009: 863)
120	Rasōighara	रसोई घर	Kitchen (Bahri 2009: 1384)
121	Rāvaṇa	रावन	Name of the chief of the demon and ruler of Lanka (who was defeated and killed by Ram in the battle for Lanka brought on by his kidnapping of Ram's wife Sita) (Mcgregor 1997: 863)
122	Sādā	सादा	1. Plain, unadorned. To be simply dressed 2. White; blank (paper); unstamped (paper) 3. Smooth; even regular 4. Pure, unmixed; unseasoned (Mcgregor 1997: 1004)
123	Sajāvaṭa	सजावट	Adoration, ornamentation, make-up (Bahri 2009: 1603)
124	Sakhā	सखा	Intimate friend, companion, chum (Bahri 2009: 1600)
125	Sakhī	सखी	Female friend, female companion (1600)
126	Samdhi	समधी	Father in law or uncle in law of one's son or daughter
127	Sāmānya-jana	सामान्यजन	1. Alike, equal 2. Shared by others: general, universal, general education 3. Ordinary; normal, usual (Mcgregor 1997: 1007)
128	Samudramanthan	समुद्रमंथन	Churning of the ocean (Bahri 2009: 1627)
129	Sārasa	सरस	Juicy, tasty, delicious (Bahri 2009: 1631)

130	Sarasvatī	सरस्वती	Name of the goddess of speech and learning and patroness of the arts (often considered as the wife of Brahma) (Mcgregor 1997: 991)
131	Sāṛī	साड़ी	A nine yard long traditional garment worn by Indian women (Bahri 2009: 1650)
132	Śādī	शादी	Marriage
133	Śaṅkha	शंख	Shell, conch, conch shell (Bahri 2009: 1534)
134	Śaṅkara	शंकर	Peace-giving, an epithet of lord Shiva (Bahri 2009: 1534)
135	Śarmā	शर्मा	*Bāṇārasī* surname of a Barber
136	Śāstra	शास्त्र	Code of laws, scriptures, religious treaties (Bahri 2009: 1551)
137	Śatranja	शतरंज	Chess, chessboard (Bahri 2009: 1537)
138	Śilpakāra	शिल्पकार	Artisan, craftsman (Bahri 2009: 1556)
139	Śilpī	शिल्पी	Craftsman, sculptor (Bahri 2009: 1556)
140	Śiva	शिव	The good, well-being, one of the divine trinity (Bahri 2009: 1556)
141	Sindūra	सिन्दूर	Red lead, vermilion, minimum (Bahri 2009: 1661)
142	Sindorā	सिंदोरा	Vermilion case (Bahri 2009: 1662)
143	Śreṇī	श्रेणी	Class, category, rank, range, guild (Bahri 2009: 1572)
144	Śūdra	सूद्र	Member of the fourth class in Hindu social hierarchy, Dalit (Bahri 2009: 1562)
145	Suggā	सुग्गा	Parrot (Bahri 2009: 1680)
146	Suhāgin	सुहागन	A woman whose husband is living, one blessed with coverture (Bahri 2009: 1694)
147	Sūtradhāra	सूत्रधार	Stage manager (of dramatic performances) (Bahri 2009: 1697)
148	Tamāśā	तमाशा	1. Show spectacle 2. Amusement; fun; joke 3. Transfer sthg. Trivial (Mcgregor 1997: 439)
149	Ṭhākura	ठाकुर	1. Deity, God 2. Idol or image of a God (especially Vishnu) 3. God the divine being (Bahri 2009: 698)
150	Thelā	ठेला	1. Truck 2. Trolley 3. Wheel barrow (Bahri 2009: 703)
151	Tilaka	तिलक	1. Mark of sandal paste on the forehead 2. An ornament worn on the forehead 3. An eminent person of a society or family (Bahri 2009: 763)
152	Tōtā	तोता	Parrot (Bahri 2009: 779)
153	Triśūla	त्रिशूल	Trident, three-pronged spear (weapon of Shiva) (Bahri 2009: 786)
154	Ullū	उल्लु	An owl (Mcgregor 1997: 134)
155	Vāsudeva	वासुदेव	Son of Vasudeva: a title of Kṛṣṇa (Bahri 2009: 916)
156	Vāsukī	वासुकी	Name of a snake supposed to support the universe, and to have been used as a churning-rope at the gods churning of the ocean (Bahri 2009: 916)
157	Varāha	वराह	Boar, pig, hog (Bahri 2009: 1457)
158	Vidhi	विधि	Method, manner, mode (Bahri 2009: 1488)

159	Vilāsījana	विलासीजन	Luxury seeking people, wanton, debauch (Bahri 2009: 1502)
160	Viṣṇu–Lakṣmī	विष्णु–लक्ष्मी	God Vishnu the preserver of the world—goddess of wealth and prosperity and wife of lord Vishnu (Bahri 2009: 1512–1413)
161	Viśvakarma	विश्वकर्मा	Demiurge, the architect of the world, God, the architect of the gods (Bahri 2009: 1508)
162	Vrata	व्रत	Fast, pledge, vow (Bahri 2009: 1533)
163	Vṛndāvana	वृन्दावन	1. Name of the forest area near Gokul in which Kṛṣṇa spent his childhood 2. The town of Vrindavan (Mcgregor 1997: 933)
164	Yamunā	यमुना	The river Yamuna (Mcgregor 1997: 842)

References

Ashtadhyayi. Encyclopaedia Britannica [Online]. Retrieved from Encyclopaedia Britannica: http://www.britannica.com/EBchecked/topic/39646/Ashtadhyayi. Accessed 14 Jan 2014

Bahri H (2009) Hindi Shabdkosh. Neha Publishers & Distributors, New Delhi

McGregor RS (ed) (1997) The Oxford Hindi English dictionary, 4th edn. Oxford University Press, Oxford

Index

K. Patil, *Toying with the Mind*, Design Science and Innovation,
https://doi.org/10.1007/978-981-96-9227-9

MIX
Papier aus verantwortungsvollen Quellen
Paper from responsible sources
FSC® C105338

If you have any concerns about our products,
you can contact us on
ProductSafety@springernature.com

In case Publisher is established outside the EU,
the EU authorized representative is:
Springer Nature Customer Service Center GmbH
Europaplatz 3, 69115 Heidelberg, Germany

Printed by Libri Plureos GmbH
in Hamburg, Germany